AF556799

IEE COMPUTING SERIES 10

Series Editors: Dr. B. Carré, Prof. S. L. Hurst, Dr. D. A. H. Jacobs, M. Sage, Prof. I. Sommerville

INDUSTRIAL SOFTWARE TECHNOLOGY

Other volumes in this series

Volume 1 Semi-custom IC design and VLSI
P. J. Hicks (Editor)
Volume 2 Software engineering for microprocessor systems
P. G. Depledge (Editor)
Volume 3 Systems on silicon
P. B. Denyer (Editor)
Volume 4 Distributed computing systems programme
D. Duce (Editor)
Volume 5 Integrated project support environments
J. A. McDermid (Editor)
Volume 6 Software engineering '86
D. J. Barnes and P. J. Brown (Editors)
Volume 7 Software engineering environments
I. Sommerville (Editor)
Volume 8 Software engineering—the decade of change
D. Ince (Editor)
Volume 9 Computer aided tools for VLSI system design
G. Russell (Editor)

INDUSTRIAL SOFTWARE TECHNOLOGY

Edited by
R. Mitchell

Peter Peregrinus Ltd on behalf of The Institution of Electrical Engineers

Published by: Peter Peregrinus Ltd., London, United Kingdom

British Library Cataloguing in Publication Data

Industrial software technology.—(IEE computing series; 10).
1. Computer software—Development
I. Mitchell, R. II. Institution of Electrical Engineers III. Series
005.1 QA76.76.D47

ISBN 0 86341 084 7

Printed in England by Billing and Sons Ltd

Dedication

To

Agnes, James and Emma

Contents

Preface

A number of interrelated developments are changing the nature of software technology. Developments in hardware are putting greatly increased power at the disposal of the software technologist: power to use new classes of languages, such as functional and logic languages, and power to use new kinds of interface techniques, such as bit-mapped graphics and pointing devices. Developments in formal methods are providing the software technologist with the means to bring greater precision into the early stages of system development. Developments in the application of artificial intelligence research are bringing techniques such as expert systems into the domain of software technology. Developments in paradigms of system development are giving the software technologist better models of system structure and better tools and techniques for developing large systems. And developments in political recognition of the importance of technology are resulting in increased funding for research and development into software technology, through initiatives such as ESPRIT.

This book brings together twenty-one papers on these developments in software technology selected from papers presented at recent European Seminars on Industrial Software Technology. The first paper compares the directions of software technology research in Europe, under ESPRIT, and Japan, under the more sharply focused Fifth Generation Programme. The next six papers address the relationship between artificial intelligence and software technology, and give examples of artificial intelligence tools and techniques, particularly expert systems, that are finding increasing application in software technology. Developments in formal methods are covered by three papers examining the effectiveness of such methods in industry, and giving examples of how formal methods research is yielding not just improved tools and techniques but also new ways of thinking about software development. Four papers directly address environments and tools, looking both at example environments and tools and at more general issues of standards for today and the impact of hardware developments on the shape of future working environments. The next two papers are related by a concern for the human computer interface, the first reviewing recent research in this field under ESPRIT and the second describing a sophisticated yet easy to use interface to programming tools. The book concludes with five papers on system development, covering the management of large projects, reusability via parameterisation, examples of models of system structure, and fault tolerance.

The 1985 and 1986 European Seminars on Industrial Software Technology, at which the papers in this book were presented, brought together speakers working in the forefront of developments in software technology and audiences drawn from organisations in the industrial computer and related fields. The Seminars were organised by the European Workshop on Industrial Computer Systems (EWICS), whose aim is to promote the effective use of industrial computer systems through education, dissemination of information, and the development of guidelines and standards. Organising the annual European Seminars on Industrial Software Engineering is one important means by which EWICS achieves its aim. EWICS is the European branch of the International Purdue Workshop and also operates as IFIP/TC5.4 (Common and/or Standardised Hardware and Software Techniques). Its members are drawn from a wide spectrum of industrial companies, industrial research groups, university research groups, licensing agents, and software and system companies.

R.J.M.
Hatfield
January 1987

Contrasting approaches to research into information technology: Europe and Japan

D. G. Morgan

Research Director (Software), Plessey Electronic Systems Research Ltd., Roke Manor

The views set out in this paper are my personal opinions and should not be taken as representing an official view held either by the Plessey Company plc, or by the CEC.

It is now common knowledge that the announcement of the Japanese Fifth Generation Computing Programme in 1982 led to the announcement of a number of national and international programmes in Information Technology, including the Esprit programme. These are popularly considered as being in response to the Japanese initiative. However, the Esprit and the UK Alvey programmes had been preceded by some years of discussion between industrial and government representatives. They, therefore, overlap rather than follow the Japanese Programme and tend to cover a much wider range of subjects. Other differences in motivation, funding and industry participation mean that comparisons between the two programmes are difficult. However, the aims of the programmes are similar, to improve the national competitiveness in Information Technology, so such comparisons must be made.

Esprit software technology programme 1985

What follows describes the work which was undertaken by the Software Technology Advisory Panel in changing the shape of the Esprit Software Technology Programme in preparation for bids to be received during 1985.

The situation before 1985

Why should there be a need to restructure the Software Technology Programme for 1985? The Esprit Software Technology Programme had been in place for two years and a considerable response had been received both to pilot projects and to the first call for proposals. However, despite substantial numbers of applications, the overall response to the first call for proposals was seen as disappointing both by the Technical Panel and by the Commission. Disappointing both in quality and quantity.

When the Esprit project was originally conceived, Software Technology was seen as being one of the major enabling technologies required by the Information Technology business. On all sides people complain about the lack of productivity in software development and the lack of availability of skilled staff. Consequently when the programme was formulated it was allocated funds similar to those allocated to the other major areas in the overall Esprit Programme. When the call for proposals went out, it was made clear that the Commission expected an enthusiastic response.

In formulating the shape of the total Esprit programme the Commission had taken the advice of an advisory panel (which contained representatives of many of Europe's leading electronics companies) that 75% of the projects should be large scale projects; that is of greater than ten million ECUs. It was pointed out strongly by the Software Technology Panel of the day that this was inappropriate to the state-of-the-art of Software Technology and the sort of projects that were judged to be required in order to advance this field. However, the Commission's view was that the proportion of large scale projects applied to the Esprit Programme overall and was not expected, necessarily, to apply to each sub-programme. In principle therefore, this would leave scope for a number of smaller projects in Software Technology. In the event the evaluation team the Commission established rejected more than 50% of the proposals received in the field of Software Technology. Sixteen projects were placed, four projects were, after reconsideration, reassessed as being suitable for support providing that their scope was reduced and these projects were asked to re-submit. In consequence the budget allocated to Software Technology was not completely allocated.

It is worth examining why there was this rejection in a programme that had received such major publicity and in which Software Technology was seen as of critical importance. There are many explanations and the following list is a personal selection of what I think were the main problems:

A. It was felt that the Assessment Panel had applied rather too academic a standard to the evaluation of the proposals. However, the Commission had an independent enquiry made into the performance of the Assessment Panel and satisfied itself that it had in fact exercised its remit fairly.

B. It has become clear that many smaller companies did not feel that they were able to put in the investment needed in order to enter a successful Esprit bid, and were discouraged by the thought of mounting a ten million ECU project of which they and their partners had to find 50% funding. There were a number of smaller companies prepared to join consortia who were unable to find partners from the larger companies.

C. Many proposals that were submitted were badly written. They failed to meet the most elementary requirements of a proposal in not having clearly identified objectives, timetables or evaluation criteria.

D. The published programme in the call for proposals had been structured into a number of different research areas and bidders were expected to indicate the area into which their proposal fitted. It was a comment of the evaluation team that many proposals did not seem to fit clearly into one area or another and they attributed this to the non-specific nature of the published Software Technology programme in the call for proposals in 1984. It is worth commenting here that this lack of specificity had been a deliberate policy of the Software Technology Panel. It had been felt that, in such a rapidly growing field, to impose too rigid a view as to what research programmes should be undertaken was to curtail the inventiveness of proposals. In the event this was probably a mistake, although at the time it seemed to be a very sensible approach. In drawing up the new 1985 Programme this was particularly addressed.

E. Industry, despite having an acknowledged crisis in this field, was not prepared to invest large sums of money in collaborative, pre-competitive research into Software Technology, either because they were not investing in Software Technology at all, or because it was felt to be too competitive a subject.

As a consequence of this poor response, in contrast to some very enthusiastic responses in other areas such as computer integrated manufacture, the Commission came under pressure from industrial representatives to consider the reallocation of money within the overall Esprit programme. The argument being that clearly there was not the interest in developing the field of software technology. It is to the Commission's credit that they resisted such pressure and in fact encouraged the technical advisory panel to create a new programme which gave the opportunity to bidders to catch up with the lack of support they had given in the first year.

Preparing the 1985 programme

When the Technical Panel met again to start the consideration of the programme for 1985, the facts given above were presented to them and they were asked to reconsider what shape the programme should have in order to achieve a more satisfactory response in 1985. In discussions which followed at Esprit Technical Week and other occasions, it became very clear that there were two main criticisms of the existing Software Technology Programme.

1. The published programme was thought to be too vague and was too sub-divided and therefore appeared more complex than was intended. It was difficult for proposers to focus on one aspect of the programme.

2. Rather than needing to push the frontiers of Software Technology forward, the real problems facing Software Technology were that there was insufficient practice and use of existing methods and tools let alone the need to develop more advanced tools. This had not been identified in the published programme.

It is clear that on the surface this is not a particularly glamorous message to put into what is thought to be a very forward looking research programme. However, it is a commonly voiced slogan that Software Technology is about creating a new discipline of software engineering and engineering is about practical skills. Examination of the world literature in the applications of software engineering yields relatively few papers on the comparative evaluation of one technique against another in terms of the practical development of large programmes. Much of the reason for this is, of course, the great difficulty in carrying out effective evaluations. In redrafting the Software Technology Programme therefore these criticisms needed to be countered. A very large software project contains all the problems of any project which requires the co-ordination of the efforts of some hundreds of human beings with widely different talents and personality; all of whom are attempting to carry out an extremely difficult and complex task to what are usually very tight time scales. Very few projects are sufficiently alike to allow for close comparison of different methods between different projects. Much of the current drive of Software Technology is supported by an act of faith that the techniques that are being introduced will lead to very much more efficient software production. Certain trends, such as that towards formality, appear self evidently beneficial as leading to greater rigour and hence to higher quality and reduced testing time on software.

It is interesting to compare the current attitudes in many countries towards computer-aided circuit design with that towards software tools. There seems to be little question as to the desirability of software tools to aid complex circuit design, but little evidence that software tools are considered as useful, although it can be argued that the software task is more complex. The ability to measure, quantitatively, the product in the former case must be a major factor. Until more effective metrics are available, the introduction of software technology will be hampered by the inability to be justified in commercial terms.

To turn once again to the 1986 programme, after recognising these omissions of the previous programmes it was decided that a new strategy should be adopted in formulating the programme for 1985. This strategy can be summarised as follows:

1. Software Technology was agreed to be still a major enabling technology of systems design. Its main purpose will be to assist in the more rapid introduction of new products and in the reduction in total lifecycle costs.

2. Taken as a whole, Europe had an expertise in software technology that was the equal, if not superior to that in Japan and the United States. However, its problem was that the expertise was spread widely throughout Europe and the aim of the Esprit programme should be to co-ordinate this expertise and to ensure that it was adopted by industry.
3. The reasons for the lack of adoption by industry of existing techniques were examined and it was felt that there were two major factors, one was the cost of implementation of many of the modern techniques and secondly, the lack of information available to middle management as to the benefits to accrue from these techniques. That is, the Project Managers within organisations who, faced with the need to set up a project team for a new project will, in general, use those techniques with which they are familiar and whose benefits have already been proven. The lack of such information at that level would obviously be a major barrier to the adoption of new techniques.
4. Consequently, it was decided that a new class of projects should be brought in to introduce software technology and to evaluate and disseminate the effectiveness of the technology.

A programme was therefore re-cast:

1. To provide a more precise statement of the desired projects.
2. To fit in with projects already placed including pilot projects.
3. To alter the balance of projects towards supporting the adoption of software engineering by industry.

This gave rise to a programme of the following shape:

The old matrix structure was thought to be too complex and was simplified so that three areas were retained and a fourth introduced:

1. Theories, methods, tools.
2. Management and industrial aspects.
3. Common environment.
4. The concept of demonstrator projects which would allow for the benefits of this technology to be demonstrated.

In reviewing this programme it was felt that the basic development of Support Environments and Tools were well covered by the SPMMS Project and the PCTE Project. However, it was felt that new work was required in:

1. Integration of hardware and software design.
2. Alternative and complementary methods of software development (in particular the co-ordination of the artificial intelligence work which was tending to use functional and logical languages).

3. Software engineering for small highly critical software (it was felt that although most of the attention was being devoted towards large scale software production there were equally critical areas where a piece of software, not in itself very large, would however form a very critical part of a piece of equipment – for example an ignition system for a motor car).
4. Metrics for software and for methodologies (this was intended to be part of the greater examination of the evaluation of software problems).
5. The man-machine interface problem for tools and environment (this was seen as being an attack on the major problems of the cost of the introduction of many of these techniques).
6. Projects covering a wider application area.
7. The evaluation projects referred to above.

The shape of this programme was welcomed by the Commission as being a more hard headed approach to the problems of software engineering and the programme was given initial acceptance. However, a subsequent visit to the Japanese Fifth Generation Conference in November 1984 reinforced the Commission's desire to introduce, even more strongly, the theme of demonstration projects.

They were very impressed with the clear cut goals of the Japanese Fifth Generation Programme which had been set at the outset and which had been adhered to. This contrasted with the very wide ranging aspirations of the Esprit Programme which covered a very much wider range of activities than that of the Japanese. It was felt that the Esprit Software Engineering and the AIP Programmes would benefit from an even more focused approach in the future and so efforts were made in the first few months of 1985 to draw up a further refinement of the aims of the Software Technology Programme and, in particular encourage industry to get together in larger groupings to form large scale demonstrator projects. In the preliminary discussions, however, it became clear that many of the large companies who had been enthusiastic supporters of the programme in its early years were now seeing themselves as being faced with a resource shortage when faced with the need to provide more support for new, very large co-operative Software Technology Programmes, particularly the absence of the benefits referred to above. On the other hand, nobody seemed to have found the secret of encouraging smaller companies to join, although various ideas had been suggested. For example, a large industrial organisation should act as the focal point for a number of smaller proposals which would be gathered together under the management of the larger organisation.

At the same time attempts were made to identify quantitative goals for the Software Technology Programme. The Japanese had announced at the Fifth Generation Conference that they were considering starting a software technology programme with the declared aim of changing the degree of automation in the process of software development from 8% to 80%. Attempts were made, in a series of special meetings of technical representatives of major electronics and telecommunications companies, to draw up a similar set of goals for the Esprit Programme. Some reluctance was shown by representatives based upon:

(*a*) The lack of true measurements of current techniques against which to compare improvements.

(*b*) The usefulness of such simplified criteria in the current state of software development.

(*c*) The difficulty of adding such criteria to a programme now well advanced, which had not been started with these criteria in mind.

However, some targets were suggested although these have not yet been officially published.

Further, it was considered that, in the major projects PCTE, SPMMS, GRASPIN, together with a new project to build tools backed upon PCTE, the Esprit project had a major initiative in the mainstream of modern thinking on the future of software technology.

On the surface there was much to be said for this line now being taken by the Commission, although it was really rather too late to alter the overall direction of the Software Technology Programme for 1985. Only when the results of this year's bids are revealed will we be able to see whether in fact this initiative to encourage industrial grouping to form major strategic projects has been successful.

The Japanese Fifth Generation Conference in 1984

What follows is a personal impression of the Japanese Fifth Generation Conference amplified by impressions of a number of visits made to Electronic Industry Research Labs during the previous week.

History will show, I suspect, that the Japanese Fifth Generation initiative has had a very major effect upon the interest in Information Technology in the western world and possibly, in the world at large. This interest was shown by the hundred per cent oversubscription to the conference which was held in November of last year in Tokyo. The Japanese openness in publicising the aims of their project led to great eagerness amongst delegates to see just how far the Japanese have progressed along their chosen path. However, in the two years that the programme has been running, the world's press have, it would appear, managed to embellish the aspirations of the Japanese project with the impression that the Japanese were making a determined attempt upon the Artificial Intelligence problem. I was not present at the first conference on the

Fifth Generation project but I am assured by those who were, that the impression was given at that time that the Japanese certainly intended to produce 'Thinking Computers'. It was very noticeable, however, in the opening speeches of the Conference both by Dr. Fuchi and Prof. Moko-oka that they were eager to dispel any ideas that the Japanese had attempted in the last two years to make any attack in this direction. In fact, Dr. Fuchi went so far as to say that they were not tackling the Artificial Intelligence problem but they were preparing themselves to generate hardware and software which would be the next generation of the way in which Information Technology was implemented.

As the conference progressed many official speakers stood up and repeated that theme and stressed that the attack on the Artificial Intelligence problem would come as a result of international co-operation and that there remained many years' work investigating the application area of the Fifth Generation Technology that the Japanese were developing before anything approaching Artificial Intelligence would be seen.

What then have the Japanese achieved in the two full years that the project has been underway? First of all one has the general impression that they have apparently achieved all their hardware targets for the first stage and I will be talking about those a little later on. They appear to be particularly strong in hardware design and in the operating systems software and they have developed a number of products which are of a commercial standard and of wide applicability in their own right.

It was very noticeable that they have a new generation of engineers widely read in the current literature. Comments were made by more than one US researcher that the US AI community would be hard-pushed to match the number of young post-graduates who were presenting papers during the Conference. It is now probably well understood that the Fifth Generation Programme is run by a central organisation known as ICOT to which companies and state research labs have contributed staff who work together under a director, Dr. Fuchi, towards achieving the goals of the Fifth Generation Programme. What is perhaps not quite so widely understood is that nearly every company that contributes staff to the central project, has also got in-house research programmes which parallel Fifth Generation Projects. Each company's programme may not be of such wide-ranging applicability as the Fifth Generation Project, but in general will be a sub-set of those activities which that particular company feels is relevant to its commercial future. Since the major Japanese companies are intensely competitive, it is quite likely that there are three or four identical developments going on within Japanese industry. These are not just replications of the ICOT programme, but represent an extension and exploitation of the ICOT programme and build upon the experience being obtained by the engineers contributed to that central team. So, for example, while the personal sequential inference machine being produced by Mitsubishi for ICOT has a performance of 100K

Lips Mitsubishi are producing a similar machine with voice response, image understanding and possibly faster performance. At the same time, independently of ICOT, a further programme is being undertaken by the Nippon Telegraph and Telephone Company, the state owned PTT, which has a programme underway which is probably larger than the Fifth Generation itself.

The total picture therefore is of a well focused project which is achieving its goals but because of the infrastructure of industry and research within Japan is also creating a very wide and deep understanding of the problems of developing these sort of systems and is providing a very large industrial base on which any future developments can be placed. It is not at all clear that similar strength and depth is being created either in Europe or in the United States. Overall therefore one can say that the conference was an impressive statement of efficiency of the Japanese industrial machine.

What are the specific achievements? Fig. 1 – an oft quoted diagram, shows the way in which they intend to develop their system. They are building a hardware base consisting of a relational database mechanism, an inference

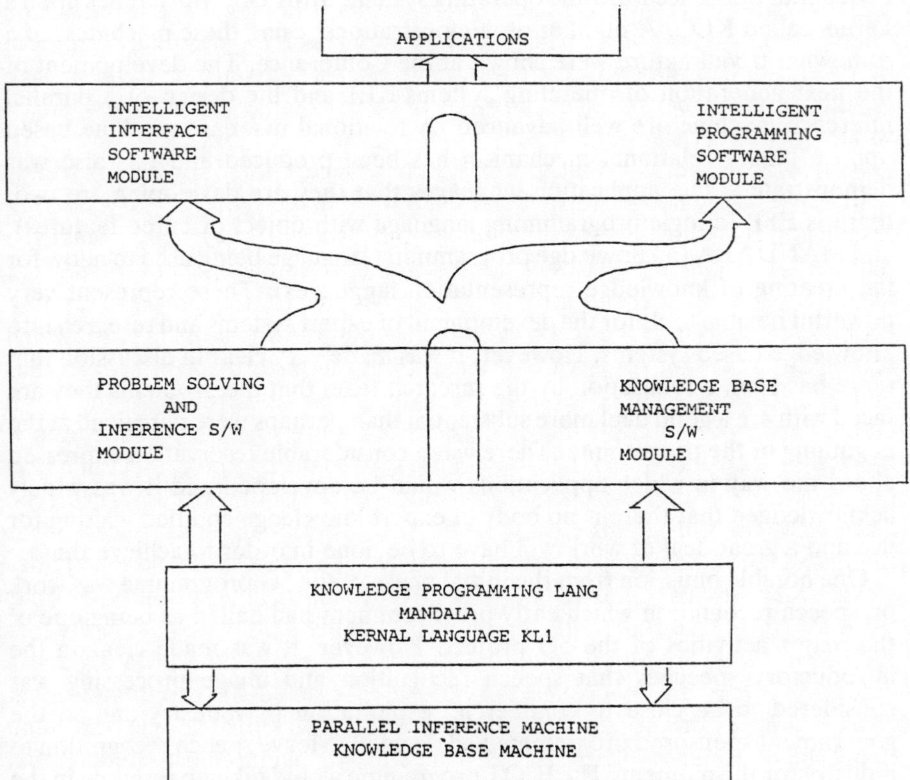

Fig. 1. *Basic software system*

mechanism, sophisticated interface hardware, all of which are linked together in an inference machine, of which there are going to be two types, firstly a sequential inference machine and eventually a parallel inference machine. Much has been said in the press about the use of Prolog, but it was very carefully stressed that Prolog has been adopted as the operating system language for their hardware and that they have not committed themselves to using Prolog alone for the implementation of knowledge based systems, in fact many of the technical papers given at the Conference were concerned with extending the concepts of Prolog to include other paradigms such as those embedded in 'Small Talk' or in 'Lisp' and one delightful experimental system called 'Tau' was demonstrated at NTT which allows for the switching of the three paradigms at will by the operator. It was a considerable tour de force by the designer but it wasn't at all clear whether the average user would be able to keep track of the complexities that would be generated in having such a wide range of flexible approaches.

All the basic equipment promised for the first phase has been produced. I will not go into details of such machines here because they have been widely publicised elsewhere. However, in summary, these are: The Personal Sequential Inference Machine operating system, SIMPOS, which relies upon a kernel called KLO. A number of demonstrations, using these machines, of a somewhat trivial nature were shown at the Conference. The development of the next generation of operating systems-KL1 and the design of a parallel inference machine are well advanced. A relational database machine based upon a binary relational mechanism has been produced and this also was demonstrated. The application languages that they are developing are two, there is ESP (a logic programming language with object oriented features), and MANDALA (a knowledge programming language being used to allow for the creating of knowledge representation languages). These represent very powerful flexible tools for the development of expert systems and research into knowledge based systems. However, it was made very clear in discussion that there has been a realisation by the research team that the problems they are faced with are a good deal more substantial than perhaps were perceived at the beginning of the programme. There was a considerable reservation expressed about the way in which applications would be developed and it was widely acknowledged that there is no body of expert knowledge codified waiting for use and a great deal of work will have to be done in order to achieve that.

One notable omission from the initial phase of the 5G programme was work on speech recognition which early press comment had hailed as being one of the major activities of the 5G project. However, it was made clear in the introductory speeches that speech recognition and image processing was considered to be close to commercial exploitation by industry and so the government sponsored programme had decided to leave speech recognition to industry for the moment. The ICOT programme would take up research in the intermediate phase. In visiting various companies, all of whom were eager to demonstrate their speech recognition, those shown were of relatively limited capability. The impression given was that most firms intended to put into the commercial market place these machines of modest ability with a view to opening up a range of applications. Further research would then be done to enhance the capabilities of these machines. However, having said that, there was a very considerable body of work evident on simultaneous translation both between Japanese and European languages and between European languages and Japanese mainly aimed at technical literature or systems manuals. Demonstrations were given which produced very passable translations at the first attempt although inevitably some nonsense statements were made by the machine. A particularly interesting application of simultaneous translation was being undertaken at NTT where they were attempting to translate dramatic newspaper statements of violent crimes. The subject matter was chosen in order to provide simple language with very clear cut scenarios and strong dramatic and contextual changes so enabling the inference mechanisms and semantic analysis to be undertaken more easily. From the evidence seen these seem to be working remarkably well.

These machines have been developed, the software exists, the first applications are being sought, but what are the immediate future plans? The programme is said to be on course and so the general direction is already given. However, the impression was formed that the detail of the programme was not being expressed as clearly as for the first phase of the programme but nevertheless there are some very impressive proposals being put forward for this next phase. These will be summarised under separate headings:

Inference sub-systems

The parallel inference machine architecture has been designed and a hundred processor prototype for hardware simulation has already been built. The processing element and the multiple element module of the real system has also been built. A software simulation for the 1000 element processor has also been undertaken and some work has been done on interfacing it with the knowledge based machine. However, a major problem area that has still to be investigated is to what degree parallelism is required in many of the problems to be faced by the artificial intelligence community. Many papers refer to problems but such results as were shown seem to indicate that parallelism of a much more modest level may well be adequate – parallelism of up to about 16 parallel channels. This is clearly a major research area for the future.

Knowledge based sub-system

The target here is to have 100 to 1000 giga-byte capacity with a few second retrieval. They are looking at a number of knowledge based machine architectures including distributed knowledge based control mechanisms and large scale knowledge based architectures, but there are many problem areas to be studied. It was the opinion of a group of European experts who met together under the auspices of the EEC Commission to consider the outcome of the Conference that the knowledge based mechanism being attempted at the moment, while very flexible, may in fact incur very substantial overheads in carrying out practical problems and future generations of knowledge based machines may see higher levels of relational mechanisms implemented in hardware.

Basic software

The kernel language for the parallel processor of the next generation will be KL1 and so the plan is to build a processor for KL1 together with its support environment and to plan a further generation of kernel languages based upon the experience gained with KL1 and KL0. This would be called KL2. On the problem solving and inference mechanism front they intend to work on system methodology to achieve a problem parallelism of 100. This refers to the problems indicated above. They see that an increased emphasis would be placed on co-operative problems solving via multiple expert systems, a technique very similar to the blackboard type of approach to large scale information sifting and analysis problems. They have also expressed an interest in moving to high level artificial intelligence but from the comments made it would appear no special insight into how this should be done yet exists. However, there is a clear wish to exploit the concepts of knowlege based management and they have a programme to develop knowledge based management software which will have knowledge representation languages for specific domains and knowledge acquisition tools. Time and time again the vast amount of work still to be undertaken on knowledge acquisition and codification in order to implement problems on these machines was emphasised.

Intelligent interface software

There was a feeling that there is still a lot of work to be done on the structure of language and they are planning to develop or continue development of a semantic dictionary and semantic analysis systems, sentence analysis and composition systems, and produce pilot models of speech, graphics and image processing interactive systems. Some evidence was put forward that they also intend to tackle the intelligent programming software problem. Although with the advent of the new software engineering programme, it is not clear how much overlap there would be. But under intelligent programming software they intend to continue with the specification, description and verification system, software knowledge management systems, programme transformation, proof mechanisms, a pilot model for software design, production and maintenance systems. This last item sounded to be very similar to the SPMMS type of project already under way under the Esprit Programme. Attempts will be made to build demonstration systems to show off the power of these new techniques.

These are a very impressive list of goals and of achievements. They are not however, unique. There was a general impression that the hardware currently available is not dissimilar to that which is coming onto the market at the moment from both European and US sources. However, there is no doubt that the very wide base of technology and trained staff and the ready availability of commercial systems and existing parallel research programmes already under way in industry, represent a very powerful momentum. It is clear that if the technology alone is going to provide the breakthrough in the field of Expert Systems and Artificial Intelligence, then the Japanese will achieve that breakthrough. However, they were very eager to ask for collaboration with other national programmes and to suggest that the real problem facing us in the future is the application of these techniques to real industrial problems. In the open discussion at the end of the Conference there was some scepticism voiced by representatives who felt that there was perhaps not a need for this full range of technology for the present state of understanding of Artificial Intelligence and Expert Systems problems. However, one or two speakers – in particular Ed Fiegenbaum – demonstrated that thinking was underway in the US into very deep knowledge based systems which would require processing powers some two orders of magnitude greater than anything being contemplated either in Japan or in the rest of the world at the present time. It is clear that there are problems already formulated in front of us which will require this technology and this represents, possibly, a way in which countries which do not yet have this level of technology and feel that they cannot necessarily repeat the research, can in fact become very active partners in this programme. That is by becoming involved in the codification of the expertise which will then be implemented using the machine.

A comparison of the Esprit and Fifth Generation programmes

In conclusion, therefore, what can one say about the comparison between the Esprit and the Japanese programmes? The Esprit programme was stated to be a reaction by the European Community to the Japanese Fifth Generation initiative. However, the Esprit programme is a very much wider ranging programme, is dealing with a very different industrial infrastructure, is covering a much wider geographical disparity both of national interests and national boundaries and, because of the 50% funding concept built into the programme, cannot expect to achieve the cohesion that the Japanese programme is achieving. However both programmes are creating significant advances in their fields, both directly through the participants and indirectly by the publicity given to the programmes. The major difference is in the approach to technology transfer. The Japanese approach appears to be to set a technological standard by a centralised collaborative programme which is used by individual firms as a yardstick against which to measure the competitiveness of their own, parallel, in house project.

The Esprit approach appears to use collaboration to encourage major companies to work together to create common technologies which they may wish to exploit in the wider market place, but which they at least have 'in-house' as a result of the project. Such collaboration is seen as a precursor to wider interworking between major European companies and, possibly, eventual industry re-structuring.

The impression gained is that the Japanese approach is likely to lead to earlier product introduction in the market place but is not suited to such subjects as VLSI process technology. The Esprit approach is likely to lead to closer future working between participating companies. The approach taken in 1985 to introduce more of a demonstrator element into the Esprit programme would seem to be strengthening the potential competitiveness of European companies. Time will tell!

The relationship between software engineering and artificial intelligence

Gerhard Goos
GMD Institut für Systemtechnik

Abstract: In this paper we discuss existing and potential applications to the field of software engineering of methods and tools developed in the area of artificial intelligence. We also indicate problem areas in the field of artificial intelligence which might be resolved by software engineers. The topics include programming by searching as a basic programming paradigm, the use of rule based systems and AI-languages, applications to rapid prototyping and program transformations. Furthermore the potential use of expert systems in software engineering is investigated.

1. Introduction

Software engineering as a branch of computer science is concerned with the theory and practical methods for efficient production of reliable and correct software in time. These issues comprise on the one side managerial questions of how to organise the software production process. On the other side software engineering is concerned with methods and tools for supporting the software life cycle, starting from requirements analysis up to the final acceptance test and maintenance phase. It is well known that presently the costs for modifications and improvements of software during the maintenance phase are much too high and amount to more than 50% of the total costs. The main issues in current research in software engineering are therefore on the one side questions of how to improve the productivity of programmers and the overall quality of the resulting product; on the other side we have the question of how to reduce the maintenance costs. It has been recognised that many of the problems stem from the fact that the design process very often starts from specifications which do not adequately reflect the intentions of the customer.

Hence a specification method is required which helps to improve the results of requirements analysis. It would be even more useful when such a specification could be made executable so that the specification can be debugged by means of rapid prototyping before the design process starts.

Starting from this question interest has been created in methods of artificial intelligence amongst software engineers. Indeed, it turns out that AI-methods might help in this situation but might also have other applications in software engineering. In this paper we discuss such applications. At the same time it turns out that some methods of software engineering might be applicable in the area of artificial intelligence.

My co-workers Reinhard Budde, Peter Kursawe, Karl-Heinz Sylla, Franz Weber and Heinz Züllighoven have contributed to the ideas expressed in this paper.

2. Programming by searching as a program paradigm

One of the key issues in the design phase on all levels is the adequate breakdown of the problem in hand into sub-problems and the corresponding construction of the solution from solutions of the sub-problems. Any method for solving this question leads to a model of software construction called a *programming paradigm*. Many such methodologies have been invented and successfully applied like top-down design (abstract machine modelling), stepwise refinement, the use of models from automata theory, etc.

One methodology, *programming by searching*, has been rarely used in software engineering up to now although it has been proven very successful in artificial intelligence. Although this methodology does not necessarily lead to efficient programs it has a number of advantages:

- The programs developed according to this methodology comprise a part which might also be read as specifications of the problem.
- These specifications might be expressed in a way which is understandable also to the non-specialist.
- Initially the specifications might even be incomplete or contradictory. Such problems might be interactively resolved by user intervention or by dynamically modifying the specifications.

The basic idea of programming by searching starts from the assumption that all possible solutions are known a priori (a potentially infinite set). These possible solutions form a *state space*. The solution(s) corresponding to the actual input data is found by successively generating the elements of the state space and by testing the generated state whether it is an appropriate solution. This *generate and test* methodology is also known as the *British Museum method*. It obviously works in practice only if the state space is sufficiently small. The method, however, shows already the basic ingredients of programming by searching: An algorithm for implementing this method consists of the following:

- an initial state,
- a set of rules for generating new states from a given one,
- a control algorithm for determining the rule to be applied next,
- a target condition characterizing the desired solution.

The method may be generalized by hierarchically structuring the state space: Possible solutions are no longer generated in one step but chains of states are generated which might be intuitively thought of as constituting successive approximations to the solution. The algorithm described above basically remains unchanged. Some specializations of this method, e.g., the greedy method, are well known to the software engineer. They are characterised by the fact that the control algorithm and the rules generating new states are integrated in a specific way.

The method, also called *unidirectional search,* however, is of limited scope only. It requires that in each state the algorithm is able to determine which rule will definitely lead to the solution (if it exists at all). To remove this restriction we may consider a generalisation of the notion of state: If each new state also comprises the information about all states generated so far we may restart the search at any former state if the last generated state appears to be a dead end (or 'less promising'). We could even generate new generalised states by simultaneously applying several rules. An abstract implementation of this generalisation uses a *search tree* instead of a chain of states as indicated in Fig. 2a. The generalised state consists of that part of the search tree generated so far. The control algorithm must now pick a node in the search tree together with a rule which is applicable to that node.

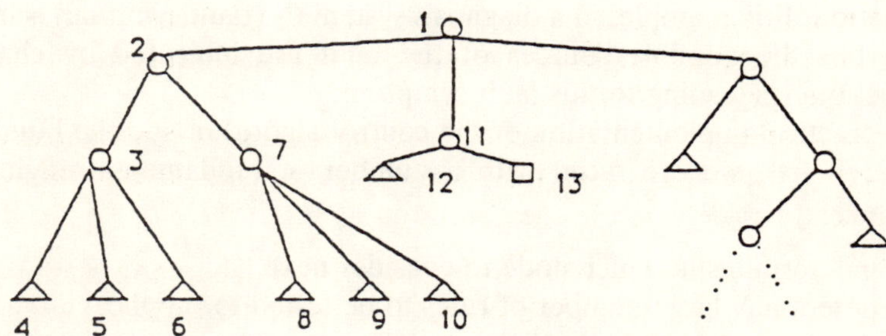

Fig. 2a *A Search Tree*

Fig. 2b shows the complete search tree for the 4 queens problem. Except for the two solutions of this problem all other leads are dead ends. Starting from the initially empty board (on level 0) there is one rule in this game indexed by integers i, k, $1 \geq i$, $k \geq 4$:

- From a node on level i–1 generate a node on level i by placing a queen in row ix, column k subject to the condition that this position is admissible (cannot be reached by any previously placed queen).

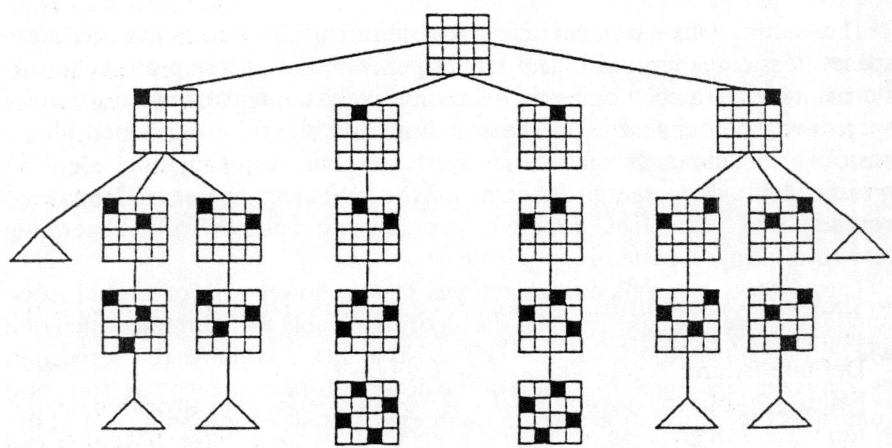

Fig. 2b *The Search Tree for the 4 Queens Problem*

The notion of search tree leads to the fact that the final generalized state also includes the path from the initial state which leads to the solution. This information is often equally useful as the solution itself: In many problems the 'solution' is known a priori but the way how to achieve it is the desired information. For example, in a diagnosis system the fault symptom is initially known but the possible sources of the fault are indicated by chains of malfunctionings leading to this fault symptom.

In a practical implementation of the control algorithm and the building of the search tree we are faced with a number of fundamental engineering problems:

- The indeterminism: which node to consider next.
- The potentially large number of rules to be tested for applicability.
- The combinatorial growth of the search tree and the associated storage problems.
- The fact that the search tree may contain potentially infinite subtrees which do not lead to a solution. (Hence the search might never terminate although a solution exists.)
- The question how to identify dead ends as early as possible.

A number of control algorithms have been developed for dealing with indeterminism: Depth-first search, breadth-first search, the use of heuristics for determining the next node and rule, etc. The book [Nilsson 1982] contains an extensive treatment of such search methods. Most of those methods can be combined with recognizing cycles which are (partially) responsible for infinite search trees. Heuristics can be applied in two different ways: Either we introduce priorities indicating the order in which branches should be added to the tree; this idea may improve the speed for finding solutions to the extent that problems become practically tractable which are beyond the accessible computing power otherwise. Or we may definitely decide that a certain branch does not lead to a solution based on heuristic arguments. Whereas the former way does never exclude solutions – at least in theory – the latter way may restrict the number of solutions being found since the heuristics might be wrong. Hence altogether programming by searching may lead to the following results:

- Success: A solution (or several/all solutions) is found together with the sequence of states leading to the solution.
- Definite failure: A solution does probably not exist.
- Failure: A solution has not been found because either the algorithm did not terminate in a given number of steps or because an unsuitable heuristics was used or because no further applications of rules are possible for other reasons.

Despite the problems mentioned programming by searching has a number of advantages compared to other programming methods:

- The strict separation of rules and control algorithm allows for considering the rules as an *algorithmic specification* of the solution process. This specification may be expressed in readable terms but is hiding all of the implementation decisions connected with the representation of the search tree and the control algorithm.
- The user may get interactively involved either for supporting the control algorithm in its decision making or for introducing state changes manually in case the system does not find an applicable rule.

Unfortunately the other side of the coin is that the algorithmic specification may be used in a restricted manner only due to shortcomings of the control algorithm.

On the other hand the interactive involvement of the user might be explored for dealing with incomplete or contradictory specifications or for extending the specifications on the fly, a direction which could never been followed in ordinary program design.

3. Rule Based Systems

Formally speaking the rules forming our specifications, often also called *productions* form a *derivation system*, $D = \langle Q,R,I,S \rangle$ where

- Q is a decidable set of states.
- p: $Q \rightarrow 2^Q$ with $p(q)$ for all $q \in Q$ being a finite (potentially empty) subset of Q. We map Q into its powerset because $q \in Q$ represents a generalised state: there may be several components to which rule p may be applied with distinct results. We write $q \rightarrow q'$ if $q' \in p(q)$.
- I and S are decidable subsets of Q representing the *initial* and *solution* states.

Derivation systems can be analysed for certain properties. The most important properties in practice are the following ones:

- A derivation system A is *noetherian* if each derivation
$$q_0 \rightarrow q_1 \rightarrow \cdots \rightarrow q_n \rightarrow \cdots$$
starting at an initial state $q_0 \in I$ terminates after a finite number of rule applications.
- A derivation system D is *confluent* if for each initial state q_0 which has a solution and for arbitrary successor states u, v of q_0 there are derivations $u \rightarrow \cdots \rightarrow z$ and $v \rightarrow \cdots \rightarrow z$, z a solution ($\in L$).

A noetherian derivation system has the interesting property that any control algorithm no matter how it proceeds will terminate after a finite number of steps because all search trees consist of a finite number of nodes only. A confluent derivation system has the property that the solution is unique if it exists at all and there are no dead ends: from each reachable state we may also reach the solution. Both properties are much desired but do not occur in practice as often as one might wish.

A control algorithm for using a derivation system $D = \langle Q,R,S,S \rangle$ is an algorithm as follows:

Initial State: $q := q_0 \in I$
  loop *select* a rule q with $p(q) \neq \phi$; exit if no such rule exists;
    select $q' \in p(q)$;
    $q := q'$; exit if $q \in S$
  end loop
Output: q if $q \in S$, no success otherwise

As discussed before it might happen that the algorithm does not terminate. The algorithm is indeterministic due to the two *select* operations. The combination of a derivation system and a control algorithm is called a *rule based system*.

Rule based systems appear in many forms in AI. Historically they have been termed *production systems*. *Frame systems* and other variations subdivide the set of rules and partially allow dynamic changes of the rule set depending on the current state.

4. AI Languages

From its beginning the people in Artificial Intelligence have used languages like LISP instead of the more common imperative languages FORTRAN, ALGOL, PASCAL, . . . It is yet difficult to analyze why these differences in language approach have occurred, notably since, by various extensions, LISP has taken over most of the features of standard languages although with different syntax. Significant features of most of the languages in AI compared to imperative languages are the following:

- Basically there are no program variables in the usual sense (although they have been mostly added through the backdoor). Computations produce new values by combining existing ones rather than by modifying them.
- Hence values like numbers from which new values cannot easily be derived by combination play a subordinate role. The basic data structures are lists of dynamic length with dynamic typing of the elements. Internally these lists correspond to binary trees. Very often lists are used for representing what is known in logic as a *term algebra*. This use occurs in its purest form in PROLOG.
- All AI languages are based on an interpretive model which allows for dynamic reinterpretation of data structures as pieces of program. Although this is a very dangerous feature, it is very helpful, e.g., in manipulating the set of rules in a rule based system.

Current programming languages can be classified as

- imperative: Operations are expressed by statements which manipulate state variables.
- applicative: Operations are applied to expressions forming a larger expression. The notion of a variable does basically not occur. Some form of the theory of recursive functions is the underlying theoretic model.
- functional: A program is a set of functional equations. These equations express relationships between values and unknowns (mathematical variables); the execution of the program has to resolve equations. Most mathematical theories can be most easily transformed into this model; but the knowledge by the program interpreter for resolving the equations is sometimes specialised.
- logic: A program consists of a number of formulas in predicate logic together with a model how new formulas may be derived from the given ones.

In practice none of these language types occurs in pure form, e.g., ALGOL 68 was a mixture of an applicative and an imperative language. Certain developments in the area of logic programming languages starting from PROLOG are especially interesting because they allow for subsuming many aspects of the applicative and functional programming style.

Of course, since all of the non-imperative languages rely on an interpreter which is currently written in software it is very difficult to achieve highly efficient programs and to judge the efficiency from the program text without knowledge of the interpreter. Hence all these language styles are interesting for the software engineer mainly for two reasons: First, some of these languages are suitable for writing specifications and programs written in such a language can be run as prototypes before the actual efficient implementation is developed. Second, programs in such languages are sometimes very concise and easy to read and write; hence it is possible to develop programs in much shorter time; problems may become solvable which otherwise could not have been attacked due to the shortage of programmers' time. We demonstrate some of these considerations using PROLOG as an example language.

PROLOG [Clocksin 1981] is a logic programming language based on the calculus of Horn-clauses. A Horn-clause is an *implication:*

$$p_1 \wedge p_2 \wedge \ldots \wedge p_n \rightarrow P_0$$

written in PROLOG as

$$p_0 :- p_1, p_2, \ldots, p_n.$$

Here we have $n \geqslant 0$; if $n = 0$ we call the clause a *fact*. The predicates (or literals) p_i may have terms as parameters as shown in the following examples. Logic variables, denoted by identifiers starting with upper case letters, may only occur as terms. Hence PROLOG essentially remains in the realm of first order predicate calculus. Certain built in predicates (*assert, call, retract*) allow for reinterpreting a term as a Horn-clause and cause second order effects useful for manipulating programs.
Simple examples of PROLOG programs are

```
human(sokrates).
mortal(X):—human(X).
```

This program allows for establishing the result

mortal(*sokrates*)

in the obvious way whereas all questions *mortal*(*xyz*)? would be answered with *no* as long as it or *human*(*xyz*) is not established as a fact.

The use for prototyping and the short way of expressing problems may be seen from the following programs for algebraic differentiation. (*dif* (*E,X,DE*) means: *DE* is the derivative of expression *E* with respect to *X*.)

```
dif (U + V,X,DU + DV):—dif (U,X,DU), dif f (V,X,DV).
dif (U*V,X,(DU*V)+(U*DV)):—dif (U,X,DU), dif (V,X,DV).
dif (X,X,1).
dif (Y,X,0):—Y≠X.
```

The query

$dif\,(a^*x + b, x, L)$

will be answered with

$L = ((0^*x) + (a^*1)) + 0$

which is the desired answer but yet not simplified.

It is obvious that this program for computing derivatives is much shorter than anything which we could write in ordinary programming languages. At the same time it is much more readable. It therefore can serve as a specification which at the same time is executable, i.e., as a prototype for a real implementation.

Horn-clauses can also be considered as the rules of a rule based system. Viewed in this way we may ask what is the control algorithm underlying the execution of these rules. In PROLOG this control algorithm is depth-first search with backtracking: For establishing the validity of a predicate the definitions (Horn-clauses) of this predicate are searched in top-down; for each definition the predicates on the RHS are considered in order from left to right and their validity is established.

The use of depth-first search is unsatisfactory in both ways (but so would be any other control algorithm): On the one side it introduces a difference between the abstract understanding of Horn-clauses as specifications and their interpretation during execution; we have to distinguish between the *declarative* and the *algorithmic* interpretation. On the other side depth-first search is just one of the possible search strategies. If we want to use another one we have to simulate it on top of depth-first search. Fortunately also in these cases the rules may be written in a way which is easy to understand and hence the properties of Horn-clauses as specifications are mostly retained.

5. Software Engineering Problems

As mentioned earlier programming by searching poses some technical problems, e.g., dealing with a large number of rules and with a large set of state variables, which are typical engineering problems. In many cases normal methods of software engineering may be applied to solve these and similar problems.

A standard method is, e.g., to factor the state space and the set of rules in a hierarchical fashion. The result is a rule based system containing rules which themselves are rule based system. Also changing the abstract representation is very often helpful in reducing the size of states.

Another class of problems arises from the fact that in practice certain properties might be expressed as parts of rules and as part of the control algorithm as well. For example, in many cases it is possible to predict subsequent rule applications (at least with a certain probability) once a certain rule has been used. Sometimes it is possible to use the rules 'backwards', i.e., if the solution is known but the sequence of steps leading to the solution is searched for we may start the search at the solution and work backwards to the initial state. Of course, this backward analysis, well known from other areas of mathematics and informatics, requires an adaptation of the rules and it is not at all clear under which circumstances it bears advantages. Also combinations of forward and backward analysis have been successfully used in practice, especially in the form of 'middle out reasoning' where it is assumed that we know some intermediate state of the solution path in advance.

Yet another form of interaction between control algorithm and rules occurs if the applicability of a rule is known only with a certain probability. This situation occurs very often in expert systems for purposes of diagnosis. There are several strategies for distributing the handling of probabilities between rules and control algorithm.

The foregoing problem may also be considered as a special case of dealing with uncertainty. Uncertainty may occur on the rule side as discussed or on the data side: The required input data can only be observed up to a certain degree of reliability or vary our time. In this case we may again use probabilistic rules or we may apply concepts of fuzzy set theory, etc.

All these questions are basically engineering problems in applying rule based systems. There exist a lot of proposals how to attack these problems – the interested reader is referred to the book [Hayes-Roth 1983] – which have been used in practice. But there does not exist a well developed and theoretically well founded basis for dealing with all these questions. Hence the practitioner will find methods by looking into existing solutions; but at the same time the scientifically interested software engineer sees a waste area of research topics.

6. Applications to Software Engineering

Rule based systems may be used in many areas of software engineering basically as parts of a *software production environment.* In most cases these applications will take the form of an *expert system*, i.e., a rule based system with additional components facilitating the acquisition of rules (the *knowledge engineering component*) and the explanation of what is going on. Basically the value of such expert systems may have three different sources which mostly appear in some combination:

- During requirements analysis (but also during later phases of software development) it might be very helpful to view system analysis as the process of acquiring rules describing the intended computational model yet without the necessary control algorithm. This idea implicitly leads to a specification language, namely any language suitable for expressing rules. Psychologically the customer is much more inclined to support the system analyst when he declares himself as a knowledge engineer, and when he does not pretend to acquire all the fuzzy details which might change in a computational model anyway, but asks for the rules governing the present situation (it goes without saying that this method is only a trick which might nevertheless uncover all the necessary details).

- Expert systems may be used for performing tasks which are more easily expressed by rule based systems than by ordinary algorithms. Typical examples for this approach are applications to prototyping or for performing program transformations on all levels, including the transformation of executable or non-executable specifications into more efficient program descriptions. For example most of the existing catalogues of standard program transformations can be quite easily put in the form of derivation systems.
- Lastly it is often easier to resolve interactive tasks within the framework of an expert system. Typical applications might include the tasks of configuration control based on large program libraries with modules in several versions and variants, support of testing from the planning stage up to actual testing phase, or tasks occurring as part of project management. In many of these applications the extensibility of the set of rules may be used for starting with a relatively 'unintelligent' system which then is gradually improved by adding new rules.

As an example for such extension techniques we might reconsider the differentiation program in section 4: This program can be immediately extended for dealing with arbitrary expressions by adding the interactive rule.

dif (*E,X,L*): – *write* ('*please input the derivative of*'), *write* (*E*),
write ('*with respect to*'), *write* (*X*, *read* (*L*).

at the very end. When it turns out that certain types of expressions occur more frequently we may then add the appropriate rules for automatically handling these cases.

It is obvious that these techniques applied either singly or continued bear a large potential of fruitful applications to software engineering tasks. These possibilities have been yet only superficially explored, e.g., by developing program transformation systems. But many more immediately useful applications remain to be written.

7. Literature

W. F. CLOCKSIN, C. S. MELLISH 'Programming in Prolog'. Springer 1981.

F. HAYES-ROTH, D. A. WATERMAN, D. B. LENAT 'Building Expert Systems'. Addison-Wesley 1983.

N. J. NILSSON 'Principles of Artificial Intelligence'. Springer 1982.

Artificial intelligence in software engineering

D. N. Shorter
Systems Designers plc

1. Introduction

Software Engineering is both a philosophy and the use of a set of methods for realising a specification in executable software. It is both a regime and a (set of) transformation process(es).

This paper discusses some ways in which AI techniques are being applied to support the Software Engineering regime and the set of methods appropriate to the particular application domain. Lastly an example is given of the use of knowledge-based components (as part of the executable software) to realise specified requirements directly.

2. AI Programming techniques

The range of problems being tackled by AI workers (usually on an experimental or concept-proving basis) has led to the development of rich tool sets. These generally include multiple languages, each with helpful facilities for some specific problem domain. Similarly various kinds of knowledge representation schema have been developed for particular classes of problem. Often such tools are used in combination to tackle various facets of the same problem. Lastly, these tool sets are supported by a powerful HCI (Human Computer Interface) and much has been learnt from developments such as Smalltalk.

A new generation of Software Engineering tools is now being developed which exploits whatever AI programming concepts are useful. In particular, much greater emphasis is being placed on the machine-assisted use of graphical notation. (One problem here is the lack of a widely accepted non-proprietary standard for high resolution bit mapped displays with mouse/window/menu interfaces, so detracting from portability and hence market size for commercial products.)

Similarly AI languages such as Lisp, Prolog and (mainly in the UK) POP-11 are coming to be used in the development of systems not necessarily having a knowledge-based component. For example, complex requirement interfaces can be expressed as Prolog rules, or the advanced pattern matching facilities of POP-11 can ease situation assessment in a command and control system. At this level AI languages are just one more set of tools available to the software engineer to address project requirements as and when they arise.

3. Knowledge based software engineering

Almost by definition, the process of software engineering demands professional methods and integrated tools which assist in the application of these. Such tools need to support not only system development but also the evolution of the system throughout the system life cycle. While some tools do exist, generally the methods which they support are hard-wired into the tools and it is difficult to modify implementations to suit particular project needs. (For example, a set of access rights, defined to enforce a general management view of how systems should be developed, may need to be enriched to meet a particular project auditing or traceability requirement.) The packaging of each method (the documentation, vocabulary, training format, HCI etc.) is generally unique and computer assisted training is rare. Lastly comprehensive help or explanation facilities are unusual and (where they do exist) are again method-specific.

There are two general issues in developing such a set of tools; the first relates to the design process itself and the second to the need to be able to evolve the designed system without losing the methods, insights and constraints encountered during the original design process.

In an assessment of the design process (Mostow '85), design is regarded as 'a process of integrating constraints imposed by the problem, the medium and the designer'. The approach suggested is 'to use rules to represent these constraints (of the application domain, or the run-time target environment and the designer's principles and methods) and to provide a mechanism by which the designer (software engineer) can steer his way to produce a design (program) which satisfies these constraints . . .'. (Parenthetical remarks are the author's interpretations for software engineering.)

The process of system evolution has similar but additional requirements. As Phillips has pointed out in his work on programming environments (Phillips 83), system evolution, that is the process of modification to meet changes in system specification, is not well supported by current design techniques which depend on interface (e.g. package) specifications and decomposition relations alone. Many constraints and relations between systems requirements and software functionality are not represented by such interfaces and there is a 'need for mechanisms for viewing and manipulating systems as relational and constraint networks'. Phillips describes a knowledge-based programming research environment (CHI) that has three modalities:

- a programming system which generates Interlisp executable code
- a vehicle for modifying knowledge about the programming process. This requires the ability to express concepts about the system directly (a self-reference capability) so these concepts can be augmented for new methods or requirements, used in explanations, etc.
- a tool for developing systems (and in particular CHI itself, enabled by its self-reference capability).

CHI includes a Very-High Level Language which extends the concept of a program 'to include programming knowledge expressed as rules, descriptions for programming concepts, declarative annotations for these, constraints and both declarative and procedural code . . .'.

Our objective at Systems Designers is to produce an assisted integrated development environment that supports well-regarded methods using concepts similar to those of CHI. A key issue is the integration of existing methods. As far as is known, there is no commercially available environment that covers all phases of the software life cycle, and the user needs to select several methods or techniques, each having different diagrammatic notations, semantics, vocabulary and even hardware support requirements. This incoherence of methods is a problem to be addressed by using a knowledge-based representation of the interfaces of method tools and providing an environment which manages the constraints of the life cycle development process. This work will build on integrated environments such as ASPECT (a development under the UK Alvey programme).

The particular application domain considered is that of the development of real-time embedded software running on a distributed target architecture. The scale of such developments will often require the management of a large project team, using a network of host computers. The initial methods supported will be selected from those frequently used for such developments. One particular need is for the method framework to be powerful enough to allow new methods to be added or existing methods to be augmented. A knowledge-based, self-referential approach is seen as essential. Hardware limitations (and commercial pressures) mean that a mix of method implementation techniques will be needed – some procedural, some declarative. Experience with mixed language environments such as POPLOG shows how this may be achieved.

4. A knowledge based project environment

Our previous work on ANALYST has already been described to EWICS (in 1984). ANALYST used knowledge-based techniques to support the rules of the CORE methodology for eliciting and representing system requirements. Development work now aims to extend this to a range of method support tools within an integrated environment as described in the previous section. The environment will also include other components such as compilers, an expert system shell and tools for rapid prototyping. Each tool installed in the environment will be accessible by a unified, object-orientated Human-Computer Interface. Configuration control tools (implementated by rule-based versioning) will control these tools and objects developed with their assistance. Tools will be involved explicitly or dynamically by situation triggered demons. Tools will produce various forms of output (text and/or diagrams) and will use a common knowledge representation format to allow interworking.

The environment itself will need to run on a variety of host configurations, including networked combinations of PCs, large minis and main frames and support to teams of developers with task allocation and controlled sharing of developed components.

Consider the input of text corresponding to some level of elaboration of application development (ranging from initial requirements expression through to package bodies). As represented in Fig. 3a, any such input will be analysed by a tool appropriate to that level of description into facts and relations (expressed in concepts for that level) together with format information to allow regeneration of the original text. Each such tool embodies a method, and the generated facts and relations express, as application rules, a fragment of the application requirement in the concepts of that method. To avoid 'representation skew', user edits may be made only to the original text, not directly to the facts, relations or format regeneration components generated after analysis.

Fig. 3b shows how application rules for a certain method are combined with rules expressing legal relations between concepts of the method, and rules

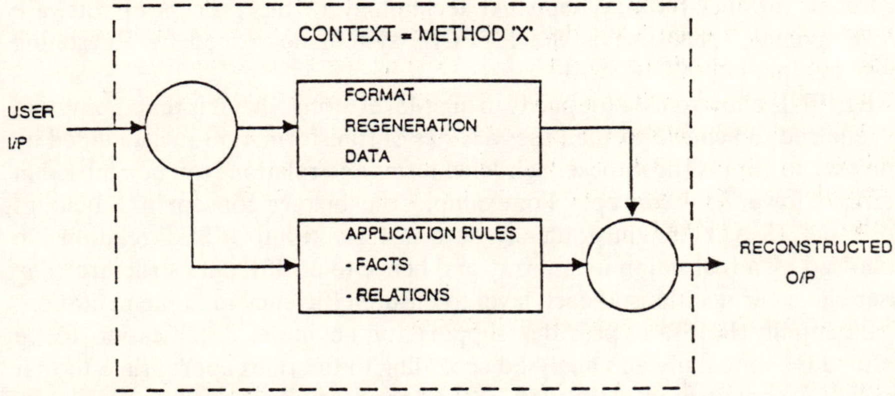

Fig. 3a *Representation of user input*

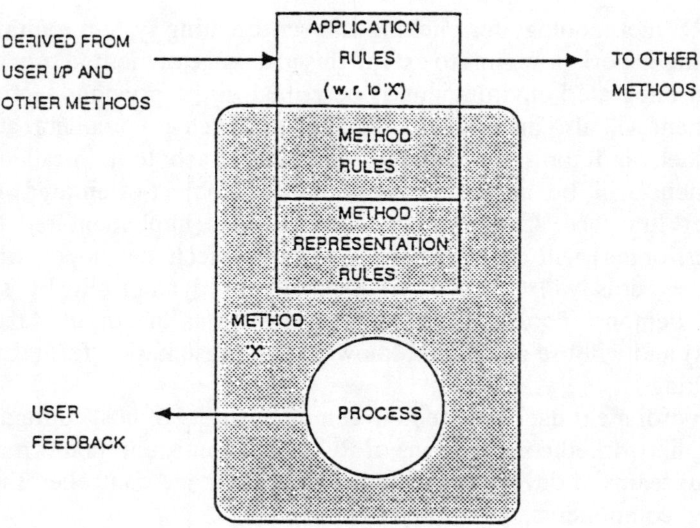

Fig. 3b *A method*

about the representation of those concepts (as text or diagrams). These rules are used by a method process to ensure correct use of the method, provide explanations, etc.

The initial set of methods chosen for the application domain of the project is shown in Fig. 4. It is anticipated that other methods and target languages will be installed to cover other application domains at a later stage.

Lastly Fig. 5 shows how the methods are combined with transition methods within the environment of life cycle control and project management methods.

The architecture sketched out above is for a development project. However there are already examples of commercially available technology (rather than software products) which apply AI technology to the process of software development. One such is the REFINE system developed by Reasoning Systems Inc.

REFINE allows a user to specify in high-level terms what it is that the system is required to achieve. It then uses a series of transformation rules, guided by the user to express how these high-level terms and relations can be realised in terms of lower level concepts. For example, the 'belong' concept in 'X belongs to Y' (X IS-A Y) might gradually be transformed into a SET relationship followed by a realisation in an array and hence to a LISP data structure. The user can steer decisions at each level to ensure efficiency in implementation.

Essentially REFINE provides support for an initial specification to be refined incrementally and analysed according to the rules appropriate to that level of decomposition. At the lowest level the specification will emerge in the target language (currently Lisp). As well as method-specific knowledge,

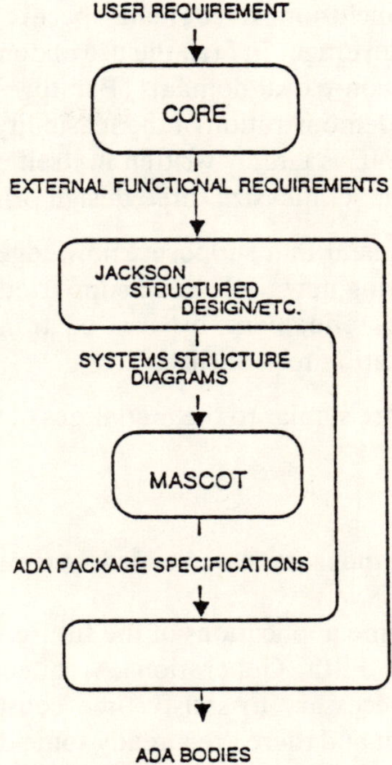

Fig. 4 *An initial method set*

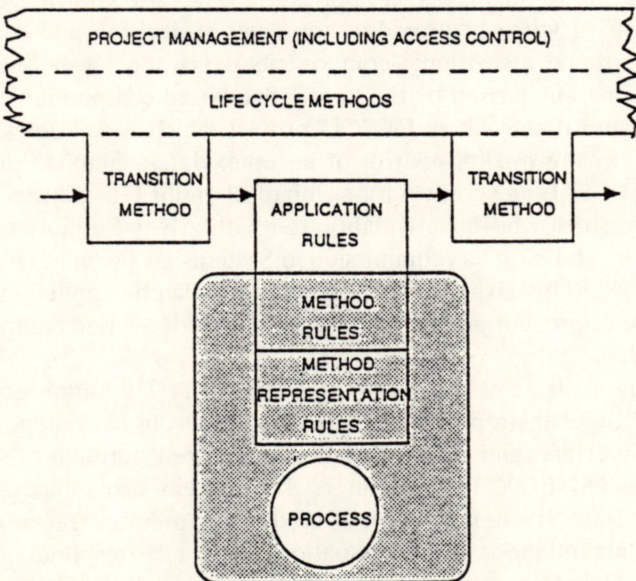

Fig. 5 *Methods in a controlled environment*

REFINE allows the inclusion of domain-specific knowledge providing increased development leverage. In fact, the use of domain specific knowledge may well be crucial for non-trivial domains (Barstow '84).

REFINE itself is one demonstration of the feasibility of this approach – it is self-describing so REFINE is largely written in itself and bootstrapped.

Reasoning Systems Inc. emphasize three design principles for REFINE:

- an integrated environment that supports knowledge-based programming
- self application, allowing new tools to be supported
- support for stepwise refinement by use of transformation rules with appropriate representation formalisms.

(These three principles are similar to the modalities of CHI described earlier.)

5. Knowledge-based components in embedded systems

Many demanding real-time applications of the future will require knowledge-based processing and 'Fifth Generation Architectures' to provide the computational power necessary to satisfy time constraints. However 'real-time' is an elastic concept and there are already some examples where benefits can be obtained by using knowledge-based components within more conventional architectures.

Such components can provide a flexibility which can be exercised (accessed) in terms relevant to the particular application. They use knowledge about the application at run-time to provide a top level of decision and control, with more conventional algorithmic components executing lower level tasks as directed and parameterised by the knowledge-based components.

One example is the Alvey RESCU project which uses knowledge-based components in the quality control of an ethoxylates chemical plant at ICI Wilton. 'RESCU' is an acronym for a club of 23 major UK industrial concerns and 3 Universities who have collaborated with Alvey Support to form the RESCU club. This club has commissioned Systems Designers to implement a demonstrator of how Expert Systems technology can be applied in real-time, and how a development project which uses knowledge-based components can be managed.

The system has been developed on a DEC VAX 11/730, running under VMS and the POPLOG environment, with a colour display used to support the plant operator's interface and plant data being acquired automatically from an already installed FOX 1/a computer. The system can be regarded as a Knowledge Base (including process modelling, process tracking, control, recipe recommendations and explanation) with a corresponding inference mechanism. Other components support the communication links to provide logged and batch data, and interfaces for both plant operators and systems developers. (Some parts of these interfaces are themselves knowledge-based – for example rules are used to steer the generation of mimic diagrams.)

In operation, starting from a set of hypotheses about the current state of the process, the plant and its instrumentation, the system loops continuously deducing consequences of that hypothesis set (i.e. tracking and modelling the plant) while there are no discrepancies between predicted and observed values. If discrepancies arise, a (rule-based) selection is made on which to consider first and a hypothesis introduced (from a pre-formed set) which might explain the discrepancies. The consequences of the hypothesis are evaluated and the hypothesis rejected if any mismatches are detected. This process continues until a new hypothesis set is found which appears to match observed plant and process behaviour and this is then used as the basis for future recommendations.

Special knowledge representation requirements (such as the need to reason with time-qualified attributes, trends, defaults, precedence, etc.) led to a decision to develop a new Knowledge Representation Language (KRL) with an associated compilation system. The representation entities produced by the compilation system are manipulated by (and in turn control) an inference process itself written in POP-11.

The project management plan for a project using knowledge based components has to take account of a different mode of development. Instead of all systems requirements being established at the early stage of the plan, a set of entities and operations upon them (sufficient to express some part of system requirements) is defined. Then knowledge representation (KRL) and manipulation structures are designed, sufficient to express this range of entities and operations. The requirements specific to the system being implemented are then expressed in the KRL and later refined during commissioning. Of course, software engineering practices are used for the specification and development of the KRL compiler, linker inferencing process and more conventional components.

Experience with RESCU has shown that knowledge based support for software engineering will need to address how some application specific requirements are transformed into executable code while other requirements are represented in a formalism which supports knowledge based processing. It has also shown the need for such support to allow the use of proprietary software components, developed outside the development project itself.

6. Summary

A development path has been outlined for an advanced software engineering environment to support an integrated set of methods both during the original design and implementation process. A number of particular requirements has been identified based on our own and other experiences. Lastly, future embedded systems will increasingly make use of knowledge based components and this needs to be addressed in designing such a development environment.

References

Barstow 84 'A perspective on automatic programming' by D. BARSTOW, *The AI Magazine*, Spring '84, Vol V, No. 1

Mostow 85 'Towards better models of the design process' by J. MOSTOW, *The AI Magazine*, Spring '85, Vol VI, No. 1

Phillips 83 'Self described programming environments' by J. PHILLIPS, Stanford University Department of Computer Science, March '83, report STAN-CS-84-1008.

Artificial intelligence in information technology
State of the art
Future and market trends

John Favaro

1. The state of the art

The state of the art in expert systems for general applications such as oil exploration and medical diagnostics is surprisingly well-defined, considering the relative youth of the field.

Fig. 6 from [FEIG 83] shows the structure of an expert system that is already being considered generic, involving components such as a **knowledge base** and an **inference engine**. This generic structure is beginning to have its effect on the marketplace with vendors offering generic components for sale.

Do expert systems for software engineering also exhibit generic features? This is a question that has a great deal of market significance, because it will partly determine whether AI systems for Software Engineering can be built in a cost-effective way from well-understood, standard components.

In seeking an answer to this question, we need to characterise the approaches, paradigms and structures that are currently being used in the construction of AI systems for Software Engineering. This will put us in a position to see whether there are trends toward a generic structure, or whether the nature of the subject makes generic structures impossible.

To illustrate the issue, consider current efforts in the construction of data bases for Ada™ Programming Support Environments. Generic data models such as the relational model have achieved such widespread acceptance that the marketplace has produced an enormous variety of efficient, cost-effective products. Yet it is becoming increasingly unclear that the relational model can be applied to the Software Engineering area. Rather, specialised variations of the entity-relationship approach or the like are being closely examined, for which there is no counterpart in the marketplace [LYON 81].

We will try to get directly to the heart of the matter by taking a close look at two specific, representative projects being carried out in the United States and Europe.

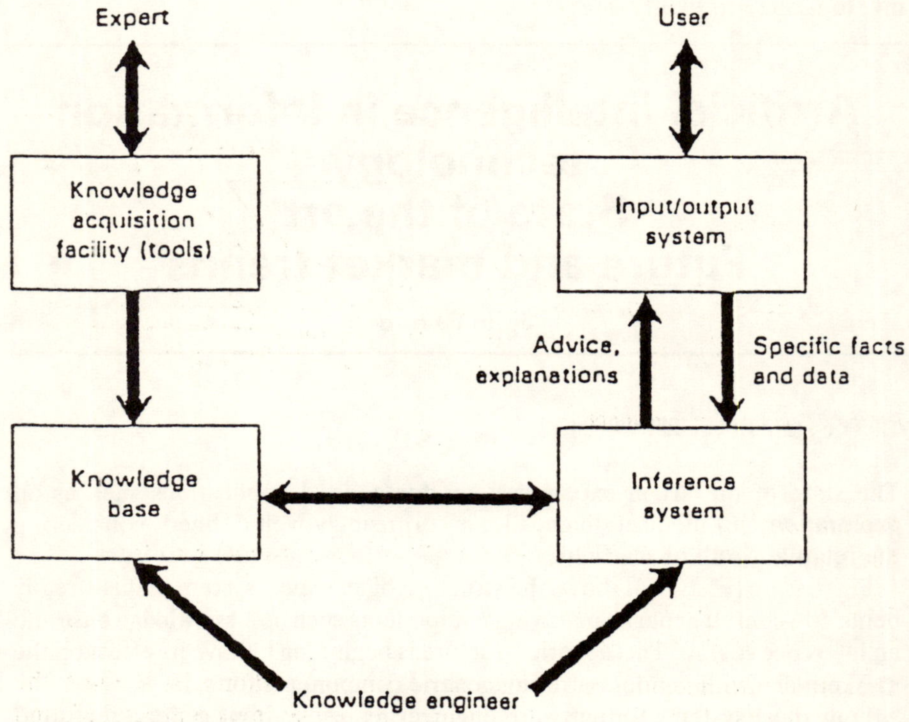

Fig. 6 *A generic expert system*

An American example: the programmer's apprentice

The Programmer's Apprentice (PA) project at MIT originated long before the current popularity of expert systems. It grew out of the proposal of Rich and Strobe in 1974 [RICH 74] into a large project that currently occupies two principal investigators (Rich and Waters), four graduate students, two undergraduates and one visiting researcher.

The goals of the PA are exactly the topic of this conference. It has become the flagship for subsequent projects involving research into AI techniques in Software Engineering and thus bears close examination.

From the beginning, the principal investigators intended to perform research into the theoretical **foundations** of knowledge-based programming. Nevertheless, they recognised early that, especially given the pragmatic aspect of Software Engineering, the community would not accept mere theories. For this reason a second line of investigation exists for the development of **demonstration** systems, both to test the theories and to experiment with how a

PA might assist a programmer. Let us examine the main results to date of each of these two lines of investigation. As Fig. 7 shows, these two lines of investigation are tightly interwoven.

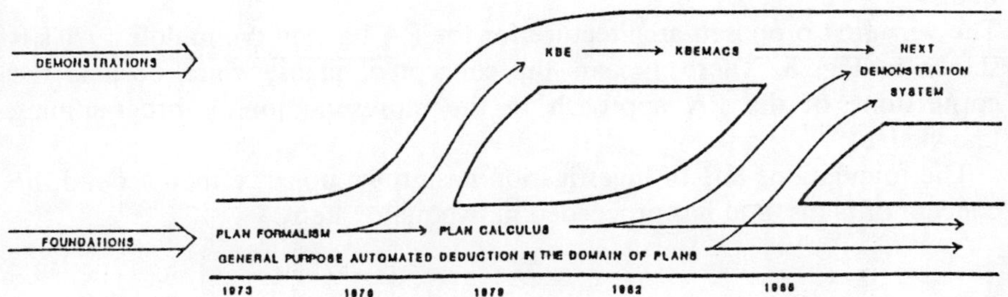

Fig. 7 *Two lines of investigation*

Foundations

The foundations line of investigation seeks solutions to two problems:

- knowledge representation
- reasoning methods

Thus they are seeking to answer the original question: How must the 'knowledge base' of a Software Engineering expert system be structured, and what is the nature of the reasoning methods contained in the 'inference engine'? We will consider the issue of knowledge representation first.

Knowledge representation

The issue of knowledge representation is actually two issues. We not only want to know how to represent programming knowledge; we also must identify the nature of programming knowledge itself. As an analogy in the 'traditional' expert systems field, researchers there have decided that sets of IF–THEN rules adequately capture the nature of expertise in some area such as medical diagnosis.

Clichés

The PA researchers decided that the appropriate contents of the knowledge base of an expert system would be programming **clichés**. These are standard techniques for dealing with programming tasks, such as sequential search or a floating-point epsilon test for equality.

A cliché is specified by identifying a set of **roles** that are played out during the application of the technique, such as enumeration and equality testing. The interaction among the roles is recorded in a **matrix**. Finally, any **constraints** on playing out the roles (such as enumerating past the end of an array) are specified.

Now let us see how the PA solved the problem of knowledge representation when that knowledge is a set of programming clichés.

Plans

The very first proposed architecture for the PA back in the middle 1970's is shown in Fig. 8. There appears the concept of **plans**, which became the cornerstone of the PA approach to the representation of programming knowledge.

The foundations line of investigation has grown up since then around the concept of plans, and has proceeded in two major steps.

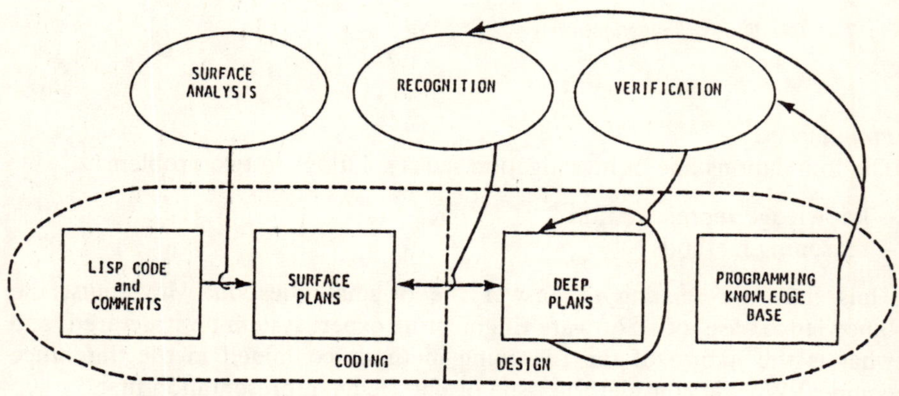

The Initial Architecture Proposed for the PA.

Fig. 8

Step 1: the plan formalism

The plan formalism was developed in the first phase of the PA project, up until 1976, pre-dating the construction of any demonstration systems. Of premier importance was the ability to represent the structure of a program in an abstract, language-independent manner.

Accordingly, a program's structure is expressed essentially as a hierarchical flowchart, where all syntactic features of programming languages are eliminated. For instance, both control flow and data flow are represented in the same way, through explicit arcs. The plan formalism contains the necessary mechanisms to represent the cliché and its roles, matrices and constraints. As an example, Fig. 9 shows the plan representation of the cliché for calculating an absolute value.

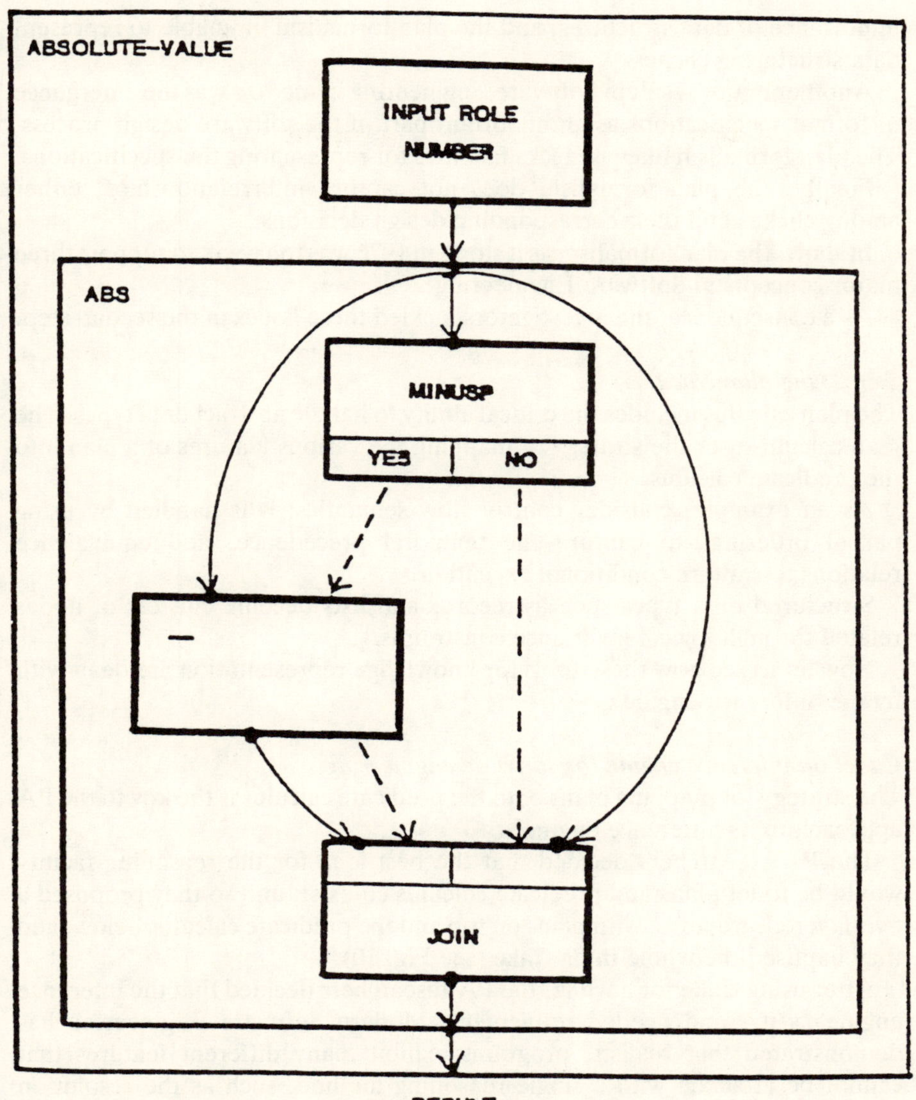

A Plan for the Cliché Absolute-Value.

Fig. 9

How successful has the plan formalism been in capturing typical programming structures? The variety is quite impressive: data flow, control, inputs, outputs, loops, recursion and subroutine calls are handled.

But these are all computational clichés. However, a major result of the Software Engineering revolution of the 70's was the recognition of the importance of data structures, and the plan formalism in unable to represent data structuring clichés.

Another major result in Software Engineering in the 70's was the emergence of formal specifications as an important part of the software design process. The plan formalism likewise lacks facilities for representing the specifications.

Finally, the plan formalism does not capture interrelationships, either among clichés and their corresponding design decisions.

In short, the plan formalism as it stood in 1976 was too weak to support three major concepts in Software Engineering.

As a consequence, the investigators tackled these issues in the second step.

Step 2: the plan calculus

The plan calculus includes the critical ability to handle abstract data types. The plan calculus uses the strategy of mapping the various features of a plan into the predicate calculus.

As an example, consider control flow semantics: it is handled by using partial orderings to capture the temporal precedence, and equivalence relations to capture conditional execution.

Structured data types such as records and lists become clusters of nodes related through logical invariance constraints.

Now let us see how these tools for knowledge representation are dealt with by the 'inference engine'.

Cake: an inference engine for software engineering

The strategy for mapping plans onto the predicate calculus is the key to the PA approach to its 'inference engine'.

The PA researchers decided that the best idea for the reasoning facility would be to let plans and predicate calculus co-exist, and so they proposed a two-layered structure, with plans on top and the predicate calculus below, and they baptised the whole thing **Cake** (see Fig. 10).

After using Cake for a while, the PA researchers decided that the inference engine as it stood needed refinement. Modern Software Engineering has demonstrated that realistic programs exhibit many different features that cannot be attacked with a single reasoning method, such as the resolution method of Prolog.

So they refined the predicate calculus level into five specialised reasoning layers, and the Plan Calculus layer into three layers for a veritable wedding cake with a total of eight layers. Let us briefly examine the five predicate calculus layers, to see what kinds of specialised reasoning the PA researchers felt are important for inference engines for Software Engineering.

The world in which evolutionary design of programs takes place is not the pristine world of facts that are either true or false, but of facts whose truth value may be unknown, or worse, may be retracted and changed as the program evolves.

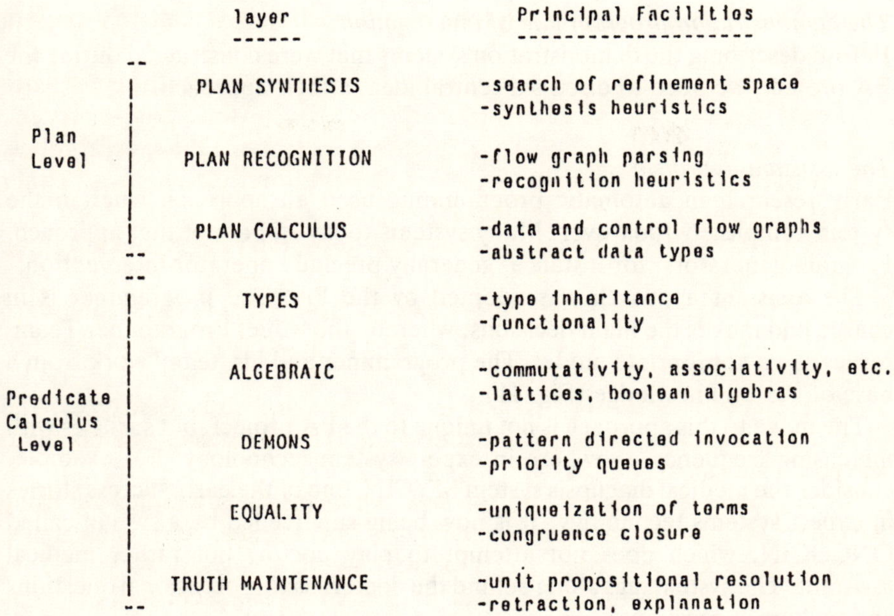

Fig. 10 *CAKE – An inference engine for Software Engineering*

Thus the bottommost layer uses concepts from [MCAL 78] to implement **truth maintenance**, whereby facts can have three possible truth values (true, false, unknown) and record dependencies among facts.

The **equality layer** allows terms to be built up out of the primitive truth values of the truth maintenance layer.

The **demon layer** makes it possible to create demons that monitor the use of terms and are triggered in order to make inferences and propagate assertions. In short, they assure a kind of 'completeness' of the knowledge base.

These three layers form the underpinnings for the real workhorse reasoning layers above them:

The **algebraic layer** contains the Software Engineering reasoning procedures, e.g. partial orders, associativities, etc.

The **types layer** adds a type hierarchy allowing reasoning about data types – certainly a crucial facility for any inference engine dealing with programming languages.

The **plan calculus** layer includes facilities for working with plans, such as querying features and altering them. Here we may see the utility of the demon layer. Since demons constantly monitor the state of the knowledge base, the mofification of a plan is likely to trigger demons to work out the consequences.

We will deal with the top two layers in a later section, since they are not yet implemented.

The demonstration program line of investigation

Before describing the demonstration systems that were constructed during the PA project, we need to discuss a central idea in their approach.

The assistant approach

Early research in automatic programming used an approach wherein the system completely took over. Many systems today still adopt that approach. Program generators, for instance, generally preclude operator intervention.

The **assistant approach** was adopted by the PA. The programmer is in charge and makes the main decisions, whereby the 'Chief Programmer Team' serves as an appropriate model. The programmer and his 'team' work from a base of shared knowledge (Fig. 11).

The move to this approach is not unique to the PA project, but surfaces with increasing frequency elsewhere in expert system technology. For example, consider the medical diagnosis system MYCIN, one of the early success stories in expert systems technology. It is now being superceded by a system called ONCOCIN, which does not attempt to play doctor, but rather medical assistant. The system tags along behind the doctor, asking occasional questions and making suggestions.

The assistant approach was a pre-requisite to the choice of a suitable domain for the construction of the demonstration systems.

Knowledge based editing

The initial phase, culminating in the plan calculus, laid the groundwork for the construction of the first demonstration system. The PA researchers settled on **program editing** as the first area to attack, since editing is intimately, unavoidably involved in the program construction process.

Whereas editors typically work on the textual structure of a program, or at best its syntactic structure, the machinery of clichés was used to allow editing in terms of algorithmic structure.

Again, this line of investigation has gone through two steps: Step 1 produced the **Knowledge-Based-Editor** (KBE) [WATE 82]. The KBE provided the first field test of the plan formalism for representing algorithmic fragments as clichés in a knowledge base.

The KBE was a standalone system. A user was a prisoner of the KBE environment. In addition, the KBE only supported LISP (not unusual in the AI world). Nevertheless, it was a useful tool for gathering early experience and some false starts could also be corrected. The researchers reported that some effort was put into providing graphical plan representation for the user according to the motto 'graphics is good', only to find that it was hopelessly confusing. The graphics were dropped.

Step 2 still used the plan formalism as the theoretical basis, but concentrated on producing a more realistic system:

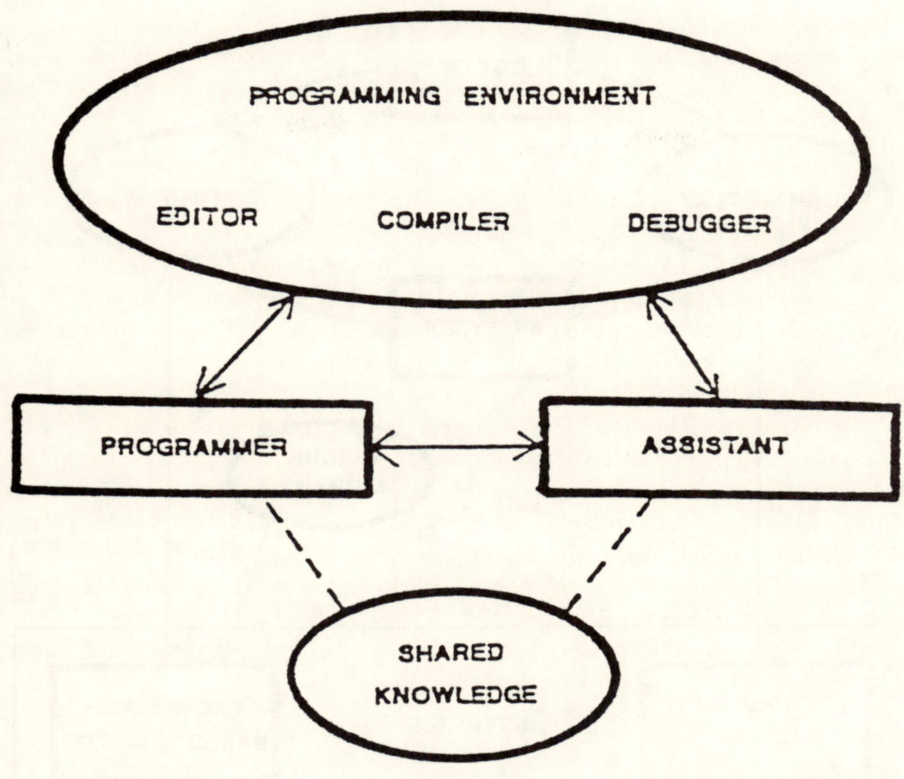

A Programming Assistant.

Fig. 11

KBEmacs

The **Knowledge-Based-Editor in Emacs (KBEmacs)** was integrated into the standard Emacs editor of a Symbolics Lisp Machine. The user, no longer a prisoner of the knowledge-based editing environment, could now move freely between algorithmic editing and traditional text or syntactical editing.

Much more attention was paid to facilities allowing the programmer to add to the knowledge base by defining new clichés.

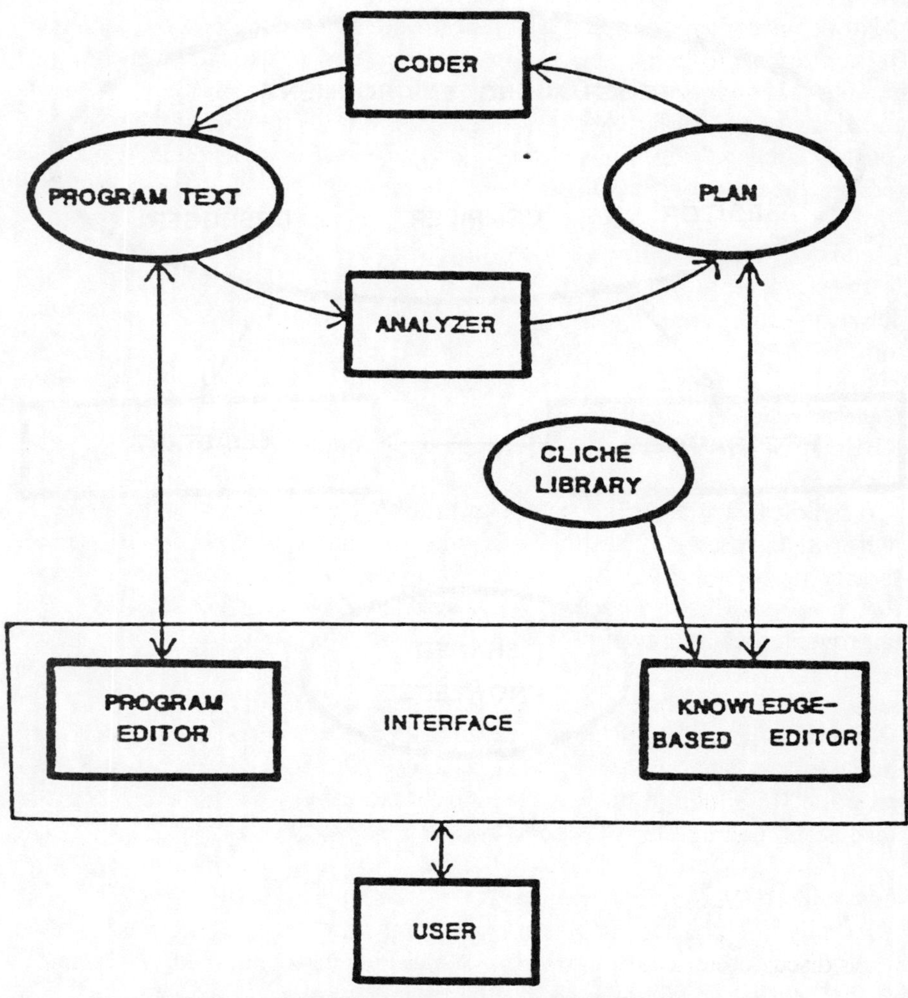

Fig. 12

And finally, in a major step toward entering the real world, KBEmacs added support for a second language: Ada. As a byproduct, this move also provided a basis for evaluating the ability of the plan formalism to represent clichés in a language-independent fashion.

The PA researchers consider it critical to preserve both text-based and knowledge-based editing capabilities, so KBEmacs maintains two concurrent representations of a program as text and plans. Knowledge-based editing is implemented as an extension to ordinary text editing, with the effect as though a human assistant were sitting at the editor modifying the text under the direction of the programmer.

Now consider what happens when the programmer modifies a program text. Its conterpart plan representation must be reconstructed. The **analyser** module takes care of this task. Unfortunately, this reconstruction process is also destructive. All knowledge of whatever clichés may have been used to construct the original plan is lost – a severe disadvantage that somewhat reduces the attractiveness of text-based editing.

The programmer's apprentice: now and the future

In this section we will assess the current status of both the foundations and demonstration lines of investigation in the PA project, and plans for the future.

Demonstration systems: now and the future

KBEmacs is written in Zetalisp and currently comprises about 40 000 lines of code.

It handles 90% of LISP. In contrast, it only handles about 50% of Ada.

Knowledge-based operations can take five minutes. In order to achieve satisfactory performance for interactive work, the PA researchers estimate that a speedup by a factor of 100 would be needed. They look to future improvements in hardware for help in this regard.

Work on KBEmacs has stopped. The researchers felt that the effort required to produce a practical system would not be worthwhile.

Work has begun on a Requirements Analysis Apprentice, as well as the next step beyond KBEmacs incorporating more elements of the original PAS proposal. In addition, pilot studies have been carried out for other domains such as debugging [SHAP 81] and program testing [CHAP 82].

Foundations: Now and the future

Currently only the first six layers of cake are implemented.

As discussed in regard to KBEmacs, the Plan Recognition Layer is needed so that text-based editing is not destructive – i.e. information about the cliché structure of a plan is not lost. Future work on this layer will be based on Brotsky's algorithm for parsing flow graphs [BROT 84]. Flow graphs as conceived by Brotsky are derived from flow grammars and serve as a model for the stepwise refinement process in Software Engineering.

Recently the researchers have begun to look at issues in reasoning with incomplete and inconsistent knowledge, motivated by problems in the acquisition and analysis of software requirements.

The programmer's apprentice: forthcoming publications

Work on the reasoning facilities of Cake continues apace. At the IJCAI Conference in August 1986 a paper will be presented by project members which will apply the ideas of simplifying assumptions (continuing previous work applied to rapid prototyping [RICH 82]) to implementing efficiently the mechanisms in Cake, i.e. typed logic and partial functions for supporting evolutionary program development [FELD 86].

Paving the way for future applications of the PA to other domains, another forthcoming paper [WATE 85] will describe an approach to translation via abstraction and re-implementation. The technique has been applied to code generation – a compiler is discussed which generates extremely efficient PDP-11 code for Pascal programs.

Finally, a forthcoming book edited by the principal researchers attests to the maturity of the project and its influence on the field [RICH 86].

Indeed, the ideas of the PA have infiltrated a number of other projects in a variety of forms. An intelligent programming tutor [SOLO 83] is based on PA concepts. An empirical studies of programming knowledge [SOLO 83] appear to confirm the cliché paradigm of programming knowledge [SOLO 84].

A European example: the knowledge-based programmer's assistant

We not turn to a project that represents one of the most significant efforts within the European Community for the application of advanced technologies to Software Engineering: The ESPRIT-sponsored **Portable Common Tool Environmnent (PCTE)**.

A subtask is currently being sponsored by ESPRIT for the development of a **Knowledge-Based Programmer's Assistant (KBPA)** within the PCTE project with GEC as the project leader [POUL 85].

The KBPA project's tasking specifies two goals: First, as research in its own right into knowledge-based Software Engineering methods, and second, as an investigation into the ability of the PCTE to support such tools.

Components of the PCTE

The second goal has a great deal of practical significance, for it represents an attempt to map the mechanisms necessary for such knowledge-based tools onto existing, well-understood components for Software Engineering support. A number of critical questions about market potential ride on the results yielded by this project.

Consequently, let us briefly review the components of the PCTE so that we may then discuss the strategy for mapping the mechanisms of the KBPA onto them.

The PCTE contains the following major subsystems:

- Basic mechanisms
- Execution mechanisms
- Communication mechanisms
- Interprocess communication
- Activities
- Distribution
- User Interface
- Object Management System

The user interface consists of a modern bit-mapped windowing environment. The Object Management System (OMS) follows the entity-relationship data model [CHEN 76].

Current Status of the PCTE

The projected finish of the PCTE project is September 1986. The final version 1.4 of the specifications will be available shortly. The C-interfaces have been specified and are available now. The Ada interfaces will follow.

The Emeraude system is the industrial version of PCTE, and is available now on field test. It will be marketed by Bull beginning at the end of 1986. Ports should begin during 1986 for a variety of machines.

The PACT project was begun in February 1986, and will last three years. The goal is to build a minimum tool set that is method independent.

A partial tool set should be available in the beginning of 1987, for use in the PACT project itself. The first public release will be after two years.

Design Strategy for the KBPA

The crucial decision in such projects is the method chosen for knowledge representation. The KBPA designers took their direction from the principles of the MIT-PA project. After considering other approaches such as the transformational approach of PSI [BARS 82], they selected the plan formalism and clichés as their theoretical basis.

Therefore, it is not surprising that the KBPA project is tackling many of the same issues as the PA project. Nor is it surprising that the structure of the KBPA is similar (see Fig. 13) with similar components:

A knowledge base contains the plans and clichés. It is administered by the Knowledge Management System.

The cliché expert is responsible for generating plan representations based upon inputs from the user interface.

The plan representation holds the program in language-independent form. It is accessed by the coding processes for producing actual target language code.

Finally, the user interface intercedes between the programmer and the cliché expert.

The Cliché Expert

The cliché expert provides an example of the pure research part not directly related to the PCTE. It has two subsystems: the design plan subsystem and searching subsystem.

The design plan subsystem constructs and maintains plans. Here again the KBPA takes its cue from the MIT-PA and uses the technique of a Truth Maintenance System [MCAL 78]. Clearly truth maintenance has become an accepted methodology for dealing with flexible evolutionary program representation with retractable truth characteristics.

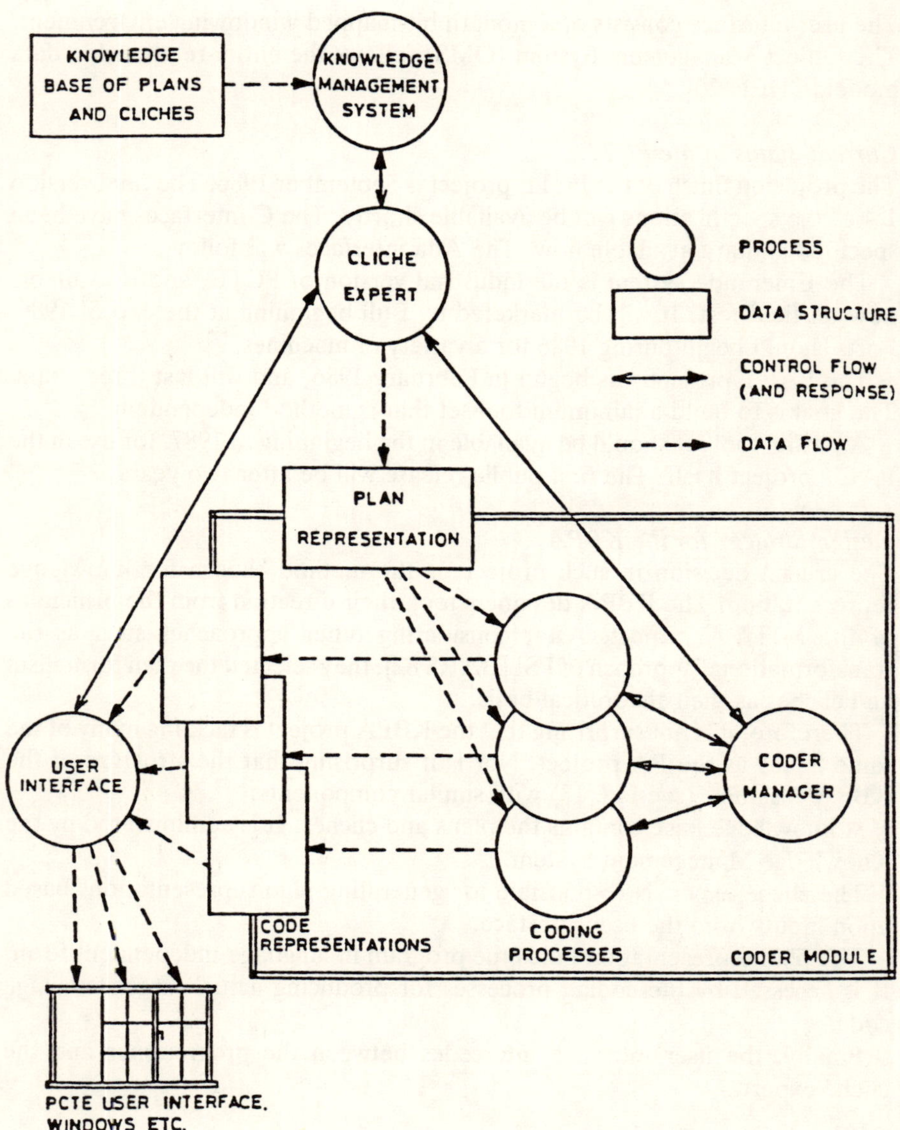

Fig. 13 *Overall design of the KBPA*

The KBPA and the PCTE

The knowledge base and user interface components provide an ideal opportunity to attempt to harness the facilities of the PCTE for knowledge-based programming. The results of this attempt will tell us a lot about where we stand in the design of versatile, powerful Software Engineering tools.

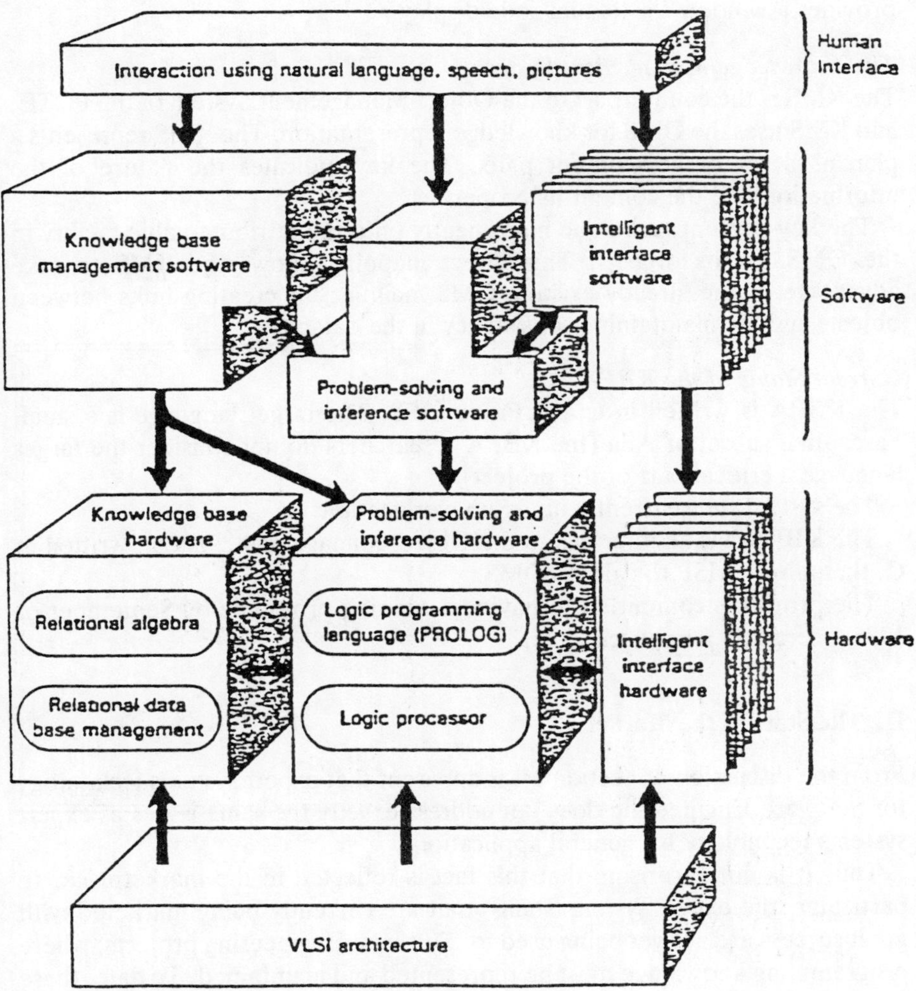

Fig. 14

User Interface
The KBPA tries to make full utilisation of the PCTE user interface, especially its windowing capabilities. The programmer is presented a number of windows, containing code, commands and other information. The mouse is used for invoking commands, together with function keys or keyboard. The

user interface acts as a 'front-end' for the cliché expert, by processing commands and their parameters as far as possible before communicating them to the cliché expert.

Since experience with the MIT-PA showed that neither the graphical nor plan representation of programs is natural for the programmer, the KBPA provides a window for textual code display.

The Knowledge Management System

The KMS is the counterpart to the Object Management System of the PCTE, and KMS uses the OMS for knowledge representation. The KMS represents a plan as a set of key-content pairs. The key indicates the nature of the information and the content its properties.

The key-content technique maps neatly onto the attribute-value facility in the OMS for its objects. This direct mapping allows the KMS to take advantage of the already-existing OMS facilities for creating links between objects and for maintaining consistency in the database.

Current Status of the KBPA

The KBPA is written in C and Franz LISP. The target language is a small Pascal-like subset of Ada (the KBPA researchers do not consider the target language a critical part of the project).

The system runs currently on a Sun workstation.

The KBPA has an interface to the Object Management System, written in C, that allows LISP to call the OMS.

The project is committed to having a running prototype by September of 1986 – certainly a tight schedule!

II. The State of the Market

From the discussion of Section I it is apparent that expert systems technology for Software Engineering does not address exactly the same issues as expert systems technology for general applications.

Thus it is not surprising that this fact is reflected in the marketplace. In particular, the expert system 'shells' that are currently being marketed with great success are not yet being used for Software Engineering projects, where programming knowledge must be represented and maintained. To date, these rule and frame-based systems have found much more popularity for the representation of other kinds of knowledge.

How can we characterize the current relationship of the marketplace to AI for Software Engineering? The current state and direction of the MIT Programmer's Apprentice project evokes two needs that must be supported by the marketplace:

- high integrated yet flexible workstation environments.
- ever greater processing power.

The trend from the Knowledge-Based Editor to KBEmacs and further indicates a desire for embedding knowledge-based programming in the surrounding Software Engineering Environment. Yet, the environment must present maximum flexibility, preferably a straightforward LISP interface, since the state of the art is still in a great deal of flux.

The PA researchers have stated that hundredfold increases in computing power are needed for acceptable performance of the Programmer's Apprentice. This need is being addressed today by the exciting new field of highly parallel computing.

Let us examine the state of the workstation market first.

Workstations for Artificial Intelligence: State of the Market

The current market for LISP-based workstations is well represented by five recognised leaders in the field.

Two of these leaders have the same origins, namely MIT where the first so-called Lisp Machine was built back in the mid-70's.

Symbolics

It is probably safe to say that Symbolics is the current market leader. Symbolics was founded in 1980 as a spinoff from MIT.

The Programmer's Apprentice is implemented on the Symbolics 3600.

The 3600 series are 36-bit workstations based on an architecture optimised to run LISP. Their predecessors were the LM-2 and the original MIT Lisp Machine.

The 3640 and 3670 have an addressable space of 1 gigabyte. The 3640 comes with 2MB of memory (expandable to 32 MB) and a 140-280MB disk drive. The 3670 comes with 168MB memory, expandable to 474MB and up to 3.8 gigabytes.

The 3640 costs around $70 000. The 3670 costs about $85 000.

LISP Machine, Inc.

Also a spinoff from MIT, they introduced their first machine in 1977. It was called CADR. The Series III CADR introduced in 1980 was the first commercial LISP machine.

Their series has been named the 'Lambda family'. These workstations connect internal printed circuit boards over the 32-bit NuBus architecture.

The Lambda 2 × 2 has two LISP processors, two keyboards and two displays. Similarly, the Lambda 4 × 4 has four parallel processors.

The series supports a multiprocessing environment. Using a 68010 co-processor with UNIX™ System V, the workstations allow simultaneous execution of LISP, PROLOG and UNIX programs.

Xerox
Xerox has long been a recognised leader in integrated workstation technology, and in fact produced one of the original integrated Software Engineering environments: Interlisp.

Xerox has focused mainly on universities, the aerospace industry, and major corporations for its AI workstations.

Two AI workstations are marketed by Xerox, the 1132 and the popular 1108. The 1108 is a single-user workstation, running the Interlisp-D system. The environment allows the framework of an application program to be reconstructed dynamically, and Interlisp allows the user to backtrack at any time during program development.

Prices for the 1108 range from \$25 000 to \$50 000, depending on configuration.

Texas Instruments
Texas Instruments has decided to invest heavily in the future of Artificial Intelligence. To that end, it has acquired 25% of Lisp Machines, Inc.

Thus it comes as no surprise that the TI Explorer workstation and the new Lambda/E workstation are nearly identical. The Explorer has a 140MB Winchester, and 60MB cartridge tapes for archival storage. It is possible to connect mass storage enclosures together to arrive at 1,120 megabytes of on-line disk storage.

The Explorer is used heavily in CAD and has a 17-inch landscape display with 1024-by-808 resolution.

The Explorer is priced from around \$50 000 to \$65 000.

Tektronix
Tektronix was chosen as a beta site in 1980 for Xerox's Smalltalk-80 system, and has been using its expertise in graphics since then in the development of bit-mapped AI workstations. In sharp contrast to the preceding companies, however, Tektronix has concentrated on the low end of the market.

Its three workstations, the 4404, 4405 and 4406, have a 640-by-480 viewable display window onto a 1024-by-1024 bit map, 1MB of RAM, 8MB of virtual memory and a 45MB disk.

The 4404 has a 68020 processor and is bundled with Smalltalk-80. LISP and PROLOG are supported. The 4405 has the faster 68020 processor with floating-point co-processor.

The 4406 has a display with 1280-by-1024 resolution, 2MB of RAM, 32MB of virtual memory address space and a 90MB disk.

The 4404, 4405 and 4406 cost around \$12 000, \$15 000 and \$24 000, respectively.

AI and the Engineering Workstation Market
The special-purpose symbolic workstations described above are not the last word in the market. On the contrary: Engineering work-stations, particularly

the Sun and Apollo series, are being used more and more for AI applications (recall that the KBPA is implemented on a Sun workstation).

Texas Instruments has recently reached agreements with both Sun and Apollo for efforts to integrate the Explorer into their environments. Sun's Network File System will be implemented on the Explorer, allowing mutual transparent file access. Similar efforts will be undertaken for the Apollo Domain environment.

Parallel Processing for Artificial Intelligence: The State of the Market

Parallel processing has become the darling of the 80's. With the rapid approach of the physical limits of processing speed, researchers see parallelism as the great hope, especially in scientific and numerical applications.

About one hundred projects are currently underway to design parallel systems, and products have begun to enter the commercial market from manufacturers such as Bolt, Beranek & Newman and Floating Point Systems.

The central techniques of AI processing, particularly searches of problem spaces, lend themselves to parallelism. Thus, expert systems technology has given a boost to the use of parallelism in non-numerical applications and many now see parallel processing as a pivotal technology for the future of AI.

Granularity in Parallel Processing

In order the establish a framework within which to discuss the state of the market in parallel processing, we introduce the concept of granularity that is gaining acceptance on the commercial market. Granularity provides a means of categorising parallel processing systems.

Fine-grained parallelism takes place at the instruction level or below. A typical example is pipelined vector processing.

Medium-grained parallelism might involve the concurrent execution of small parallelisable segments of a program such as iterative loops. This is generally a function of parallelising compilers. Inference machines for Prolog might be placed into this category.

Coarse-grained parallelism refers to the processing of tasks that may be partitioned into largely independent subtasks for separate computation.

We can get a good feel for the state of the art by examining a representative from each of these categories.

Medium-grained Parallelism: The Fifth Generation

Certainly the best publicised effort in parallel processing today is the Fifth Generation Project. Fig. 9 shows clearly the reliance of the project strategy on special-purpose, highly-parallel hardware to support the inference engines and knowledge bases.

Consider current forecasts at Hitachi's Central Research Laboratory in Tokyo. Senior researcher Horikoshi states that Hitachi's supercomputers of the 1990s are targeted to achieve processing levels of 1 trillion FLOPS, which is 1000 times faster than today's supercomputers.

Hitachi has set a performance target for logical inference machines of 1 billion logical inferences per second, or one GLIPS. One strategy under consideration is the use of 1000 processors, each capable of 1-MLIPS performance.

The Japanese projects are geared toward the specific use of Prolog and the concept of inference machines.

Fine-Grained Parallelism: The Connection Machine

As early as 1981, Danny Hillis, a graduate student at MIT, conceived an extremely fine-grained machine that would contain 64 000 single-bit processors [HILL 81]. It became known as the Connection Machine, and the Ph.D. thesis describing its architecture has been taken up into the ACM Distinguished Dissertation Series and the MIT Press Artificial Intelligence Series [HILL 85].

In 1984 at a conference on commercial uses of artificial intelligence [WINS 84], Marvin Minsky of MIT announced his intention to start a private research institute to do basic research in artificial intelligence. This intention was subsequently realised as Thinking Machines Corp. of Cambridge, MA.

In April, 1986, Thinking Machines Corp. introduced the Connection Machine as a commercial product, to considerable fanfare in the popular press.

The impressive list of company backers has attracted a great deal of attention. Some of the work leading up to the commercialisation was sponsored by the DARPA Strategic Computing Program, which invested $4.7 million and is buying two of the first six machines. William S. Paley, former chairman of CBS, Inc.; Claude Shannon, the father of statistical information theory; and Nobel-Prize-winning physicist Richard Feynman of Caltech are other prominent sponsors.

The single-bit processors are packed 16 to a silicon chip, and fit in a 1.5 meter cube. The machine can reprogram the pattern of connections among the processors to fit the problem it is working on (thus the name 'Connection Machine'). Four kilobits of memory are dedicated to each processor, and a larger memory is shared by all processors.

Speeds of 1 billion instructions per second – roughly the speed of a Cray XMP – are projected.

Presently, the Connection Machine is targeted for applications in image analysis and pattern recognition. AI applications will follow.

By early June 1986 seven computers had been ordered: MIT, DARPA and Perkin-Elmer each ordered two, and Yale University had ordered one machine.

Prices have ranged from $1 million to $3 million.

Coarse-grained Parallelism: The Hypercube

From 1978 to 1980, graduate students Sally Browning (BROW 80] and Bart Locanthi (LOCA 80] performed research on concurrent architectures at the California Institute of Technology. This research grew into the so-called Cosmic Cube project at CalTech, under the leadership of Charles Seitz as part of the Concurrent Computing program of Professor Geoffrey Fox.

The Cosmic Cube architecture has also become known as a hypercube architecture. It is a loosely coupled system of multiple processors in which each node is coupled to its immediate neighbours by high-speed, serial links (Fig. 15). The number of nodes in a hypercube is always a power of 2, called the dimensionality of the cube (Fig. 16).

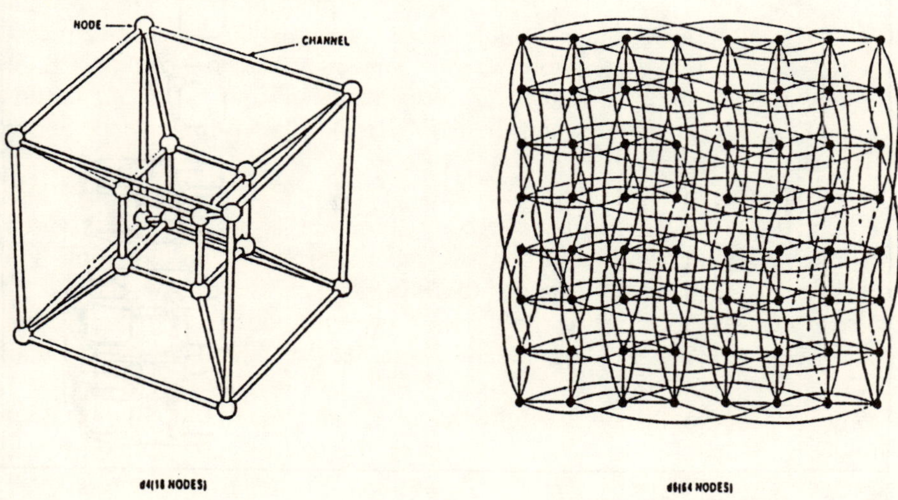

Hypercube Topology

Fig. 15 *The Hypercube Architecture*

The original CalTech prototype (the Mark I) had 64 nodes, using 8086 processors for the nodes. It went into operation in December 1983, and achieved peak performance 10 times greater than a VAX 11/780.

The Mark II of the Jet Propulsion Laboratory comes in a 128-node version, and expects performance of 25 times the VAX 11/780.

The Mark III is currently under development. The nodes will have 68020 processors, and each node will have 4MB of its own memory. A 32-node version is expected in mid-1986 and a 256-node system in 1987.

Mark III performance is projected at up to 10 000 times the VAX 11/780. The 32-node version should out-perform a Cray I, and the 256-node version should beat a Cray X/MP.

Intel Scientific Computers is marketing the leading commercial realisation of the hypercube architecture, the iSPC™ ('Intel Personal Supercomputer').

It comes in three versions: the 32-node /d5 ($50 000), the 64-node /d6 ($275 000), and the 128-node /d7 ($520 000).

There are two major components of the architecture: the cube and the cube manager (Fig. 17).

The cube itself consists of an ensemble of one, two or four computational

THE HYPERCUBE TOPOLOGY

Dimensions	Nodes	Channels
d0	1	0
d1	2	1
d2′	4	4
d3	8	12
d4	16	32
d5	32	80

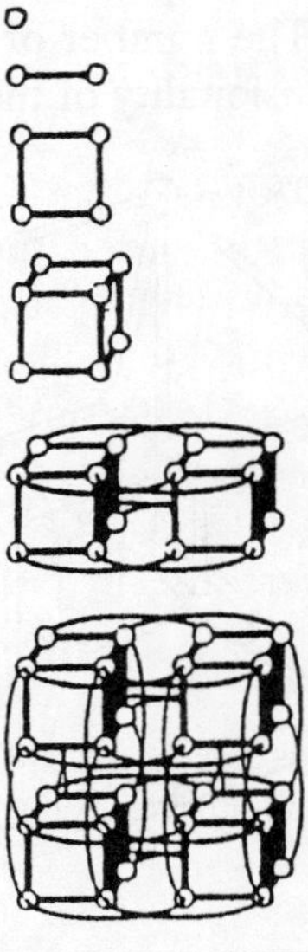

Fig. 16 *Hypercube dimensionality*

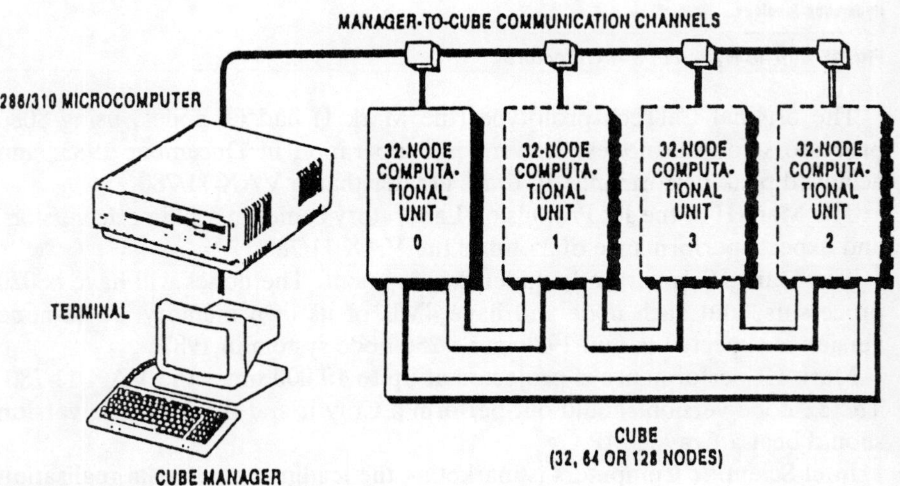

Fig. 17 *Components of the iSPC*

units, each with 32 nodes. The nodes contain Intel 80286 CPU's with 80287 arithmetic co-processors.

Internode communication is over 10 Megabit Ethernet lines, controlled at each node by 82586 Ethernet controller chips.

Resident in each node is an interrupt-driven, message-passing, kernel operating system based on the 'cosmic kernel' developed by Seitz. It supports multiple processes per node, automatic routing of messages between non-adjacent nodes, and dynamic message buffer management. Such a system has the flexibility to be programmed to handle different topologies such as ring and tree topologies.

The global Ethernet communication channel connects all nodes to the cube manager, an Intel System 286/310 computer running Xenix™ Version 3.0.

FORTRAN and C are supported in the UNIX development environment. It should be noted, however, that Gold Hill Computers of Harvard, Massachusetts plans to deliver a concurrent version of LISP for the iSPC by mid-1986 for artificial intelligence applications.

In its maximum configuration of 128 nodes, the iSPC can handle an impressive 64MB of online dynamic RAM and reach processing speeds of 100 MIPS.

There are currently 22 systems installed. Some customers have been: Yale University (/d7), Oak Ridge National Labs (/d6), NSA's Supercomputer Research Center (/d6), RIACS at NASA/AMES (/d5) and Cal Tech (/d7).

The first European installation took place in December 1985 at the Christian Mikkelsen Institute in Bergen, Norway.

THE iPSC FAMILY

	iPSC MODEL		
	d5	d6	d7
Processing Nodes	32	64	128
Memory, MBytes	16	32	64
Communication Channels	80	192	448
MIPS*	25	50	100
MFLOPS*	2	4	8
Whetstones, KWIPS*	5,750	11,500	23,000
Aggregate Memory Bandwidths (Mbytes/sec)	512	1,024	2,048
Aggregate Communication Bandwidth (Mbytes/sec)	64	128	256
Aggregate Messages (K messages/Sec)	180	360	720

*Based on 80% computational efficiency relative to message traffic overhead.

Fig. 18

Trademarks: Unix is a trademark of AT & T Bell Laboratories. Xenix is a trademark of Microsoft Corporation. Ada is a trademark of the US Dept. of Defense. iSPC is a trademark of Intel Corporation.

Appendix

Addresses of Projects and Vendors:

Charles Rich, The Programmer's Apprentice Project, The Artificial Intelligence Laboratory, Massachusetts Institute of Technology, 545 Technology Square, Rm. 839, Cambridge, MA 02139. Rich@MIT-MC

Gavin Oddy
The Knowledge-Based Programmer's Assistant Project
GEC Research Laboratories
Marconi Research Centre
West Hanningfield Road
Great Baddow
Chelmsford
Essex
CM2 8HN
United Kingdom

Symbolics
11 Cambridge Center
Cambridge, MA 02142

LISP Machine Inc.
6033 W. Century Blvd.
Los Angeles, CA 90045

Texas Instruments
P.O. Box 809063
Dallas, TX 75380-9063

Xerox
250 N. Halstead St.
Pasadena, CA 91109

Intel Scientific Computers
Director of Marketing
15201 N.W. Greenbrier Parkway
Beaverton, OR 97006

References

BARS 82 D. R. BARSTOW, R. D. DUFFEY, S. SMOLIAR, and S. VESTAL, 'An automatic programming system to support an experimental science', in *Proc. 6th Int. Conf. Software Eng.*, Sept. 1982

BROT 84 D. BROTSKY, 'An algorithm for parsing flow graphs', M. S. Thesis, Rep. MIT/AI/TR-704, March 1984

BROW 80 S. A. BROWNING, 'The tree machine: A highly concurrent computing environment', Tech. Rep. 3760:TR:80, Comp. Sci. Dept., CalTech, Pasadena, 1980

CHAP 82 D. CHAPMAN, 'A program testing assistant', *Commun. ACM*, vol. 25, no. 9, Sept. 1982.

CHEN 76 P. P. CHEN, 'The Entity-Relationship Model: Towards a Unifed View of Data', *ACM Trans. Database Systems,* vol. 1, no. 1, march 1976

FEIG 83 E. A., FEIGENBAUM and P. McCORDUCK, *The Fifth Generation: Artificial Intelligence and Japan's Computer Challenge to the World,* Reading: Addison-Wesley, 1983

FELD 86 Y. A. FELDMAN and C. RICH, 'The Use of Simplifying Assumptions to Implement and Efficient Reasoning Procedure for Partial Functions', *Proc. of the Fifth National Conference on Artificial Intelligence,* Philadelphia, PA, August, 1986. (*Submitted*)

GREE 81 C. GREEN *et al.*, 'Research on knowledge-based programming and algorithm design – 1981', Kestrel Institute, Palo Alto, CA, 1981

HILL 81 W. D. HILLIS, 'The Connection Machine', Report AIM-646, Artificial Intelligence Laboratory, Mass, Institute of Technology, Cambridge, MA, 1981.

HILL 85 W. D. HILLIS, *The Connection Machine,* Cambridge: MIT Press, 1985.

LOCA 80 B. N. LOCANTHI, 'The homogeneous machine', Tech. Rep. 3759: TR: 80, Comp. Sci. Dept., CalTech, Pasadena, 1980

LYON 81 T. G. L. LYONS, 'Examination of Relational Approach', Software Sciences Limited, Doc. No. E1719/WOR/2, December 1981

MCAL 78 D. A. McALLESTER, 'A Three Valued Truth Maintenance System', M-473, Artificial Intelligence Laboratory, MIT, May 1978

NEIG 84 J. M. NEIGHBORS, The Draco approach to constructing software from reusable components', *IEEE Trans. Software Eng.*, vol. SE-10, Sept. 1984

POUL 85 K. J. POULTER, G. C. ODDY and M. R. KENNETT, 'The Design of a Knowledge Based Programmer's Assistant', GEC Research Laboratories, 1985.

RICH 74 C. RICH and H. SHROBE, 'Understanding LISP Programs: Towards a Programmer's Apprentice', (M.S. proposal), MIT/AI/WP-82, December, 1974

RICH 82 C. RICH and R. C. WATERS, 'The Disciplined Use of Simplifying Assumptions', *Proc. of ACM SIGSOFT Second Software Engineering Symposium: Workshop on Rapid Prototyping, ACM SIGSOFT Software Engineering Notes,* Vol. 7, No. 5, December, 1982

RICH 86 C. RICH and R. C. WATERS (eds.) *Readings in Artificial Intelligence and Software Engineering*, Morgan-Kaufman, Los Altos, 1986. (*to appear*)

SHAP 81 D. SHAPIRO, 'Sniffers: A system that understands bugs,' M.S. thesis, Rep. MIT/AIM-638, June 1981

SMIT 74 B. SMITH and C. HEWITT, 'Towards a Programming Apprentice', *Proc. of Conf. on Artificial Intelligence and the Simulation of Behavior,* U. of Sussex, July 1974

SOLO 83 E. SOLOWAY et al., 'MENO-II: An intelligent program tutor', *J. Computer Based Instruction*, Vol. 10, No. 1, Summer 1983

SOLO 84 E. SOLOWAY and K. EHRLICH, 'Empirical studies of programming knowledge', *IEEE Trans. Software Eng.,* vol. SE-10, Sept. 1984.

WATE 82 R. C. WATERS, 'The programmer's apprentice: Knowledge-based program editing', *IEEE Trans. Software Eng.*, vol. SE-8, Jan. 1982

WATE 85 R. C. WATERS, 'Program Translation via Abstruction and Reimplementation', *IEEE Trans. on Software Eng.*, submitted Nov. 1985

WINS 85 P. H. WINSTON and K. A. PRENDERGAST, eds., *The AI Business: The Commercial Uses of Artificial Intelligence,* Cambridge: The MIT Press, 1984.

Development of knowledge based systems
New ways for the development of software systems

M.-J. Schachter-Radig

Artificial Intelligence and Distributed Systems, SCS Organisationsberatung und Informationstechnik GmbH, Hamburg, FRG

1. Introduction

Technological progress is characterised by the transition from a stage of extremely high creativity, in which pioneers use in an ad-hoc manner the available means to reach their goals to a structured and controlled process [Schachter 85a]. Engineering implies to have understood the production process in the sense that a suitable methodology can be derived from the experience so far. At this stage a technology can enter wide industrial use, because production plans can be created which allow risks to be minimised.

In this sense industrial and software technology are very similar, which has led to a considerable borrowing from the terminology of industrial technology for corresponding concepts in software technology. For example the term 'software engineering' emphasises that the construction of software is an engineering task. Terms such as 'software tools' and 'software factory' suggest that paradigms of industrial production are being adopted for software production [Wegner 84].

Specifically one can see similarities in the following areas. Industrial development and software development are becoming increasingly capital-intensive: their tools are becoming more powerful and expensive, and it requires greater early investment to reduce later expenditures. Reusability is a general engineering principle the importance of which derives from the desire to avoid duplication and to capture commonality in undertaking classes of inherently similar tasks. This requirement provides both an intellectual justification for research that simplifies and unifies classes of phenomena and an economic justification for developing reusable software products which make computers and programmers more productive. Compilers and operating systems are reusable resources for producing application programs in the same way that machine tools are used for the fabrication of physical goods. Machine tools can be reused for the fabrication of a large variety of products because they support atomic activities, such as grinding etc. which are common to various production processes. Alternatively a tool may perform a predetermined series of fabrication steps. Their reusability is inflexible because they have to be operated by personnel who in turn have to be controlled by the results of a previous design process, or these results are taken as basis for the numerical control of the machine tool. When a new product is required this again involves human intervention from the very beginning. Compilers, operating systems and central processor units are used in the same way in the software technology. The need to increase cost effectiveness has brought up the idea of extending the reusability of technical resources not only for common atomic activities but also for a problem class, such as the production of printed circuits rather than a specific board by use of flexible, autonomous and cooperative robots. To achieve this goal, progress in industrial technology is linked in a fruitful liaison with software technology. Computer Integrated Manufacturing is the combined technology which will require knowledge engineering to create 'machine intelligence' as outlined above.

The necessary methods and software tools are research and development topics of 'Artificial Intelligence', a discipline within computer science. H. Zemanek gives a comprehensive definition of this pretentious label: 'the term is an abbreviation for artificial generation of results which normally are produced by an intelligent mind'. This is a reformulation of the well known desire of humankind already expressed by Leibnitz and Pascal in the 18th century *the liberation of man from the burden of tedious intellectual activity* [Bauer 85].

Knowledge Based Systems are industrial products of Artificial Intelligence technology. Their engineering process, as reflected in an operational life-cycle model, is tightly connected with their intrinsic features and capabilities.

The following chapters will give a review of the following topics:
- characteristics of Knowledge Based Systems
- enabling technological environment
- classes of Knowledge Based Systems
- life-cycle paradigms for software development
- interaction between knowledge engineering and software engineering.

2. Characteristics of knowledge processing

In conventional data processing a programmer is generating a set of computer instructions, which describes the solution path for each foreseen situation which the program has to handle. The solution path is completely planned during the engineering process – any surprises in the events of processing are bugs that must be eliminated [Scown 85]. **Data processing** is calculation with predictable completion including the entire set of actions necessary to achieve the problem solution. In this case rules are part of the instruction set which describes domain knowledge. IF-THEN rules are control instructions in a conventional program which implement the algorithm for the problem solution. The algorithm is the stepwise decomposition of the solution path and describes the strategy for the problem solving. Any change in this strategy requires reprogramming.

Information processing enlarges the utilisation of computers in manipulation of numerical and non-numerical information, which is represented by complex structured data (as in data base management systems). Databases are the static description of facts which are represented by relationships between objects. The situation-dependent evaluation of facts and the problem oriented retrieval of information is the responsibility of the user or the application program. Fourth generation languages support a user friendly storage, update and retrieval of information. Communication of information between data processing systems is organised through message exchange and standardised in the OSI/ISO-protocols. The logical connection of information stored in distributed systems, for example in office systems, is solved by the document-server concept. Another example is the representation and processing of graphical information in CAD-systems. Despite the impressive variety of tasks in data and information processing which are solved by computers, in conventional software the computer is the interpreter of procedurally formulated problem solutions. The use of computers for engineers has been limited almost exclusively to algorithmic solutions such as stress analysis by finite element methods or electronic circuit simulators.

Developments in software engineering in the last decade have provided application program developers with a variety of hardware and software tools. Emphasis has been put on problems which require the fast processing of large amounts of data and the repetition of many steps. Moreover many technical engineering problems are not amenable to purely algorithmic solutions: 'The engineering method is the use of heuristics to cause the best change in a poorly understood situation within the available resources' [KOEN 85]. 'This is as well true in the highly judgemental decision tasks found in business and finance' [Hart 86]. Knowledge Based Systems provide a programming methodology for solving ill-structured problems. Since these systems also provide a flexible software development methodology – by separating the knowledge base from the inference mechanism – they are not exclusively of increasing interest to the engineering community.

Problems involving decisions, based on the complex interaction of many factors, which must be considered as a whole, rather than as a series of steps, are suitable for solution using techniques, such as **symbol processing**, which are specific to Artificial Intelligence. The symbols processed by Artificial Intelligence programs often represent real-world entities, and instead of performing calculations they manipulate relationships among the symbols.

Symbol processing is not the processing of the content of variables, rather it is the manipulation of symbols independently of their present content. This allows the preparation of a problem solution even if the value of a symbol becomes known later on. For example, a car has a colour, but only in the latter stages in the manufacturing process is it necessary to know the specific colour of a certain car. Nevertheless the fact that a car has a colour has to be known already at the production planning level, but only the spray painting station is interested in the specific colour or it becomes relevant at customer delivery. In conventional software each variable has to be instantiated from the very beginning, each variable has to have a value. Symbol processing of Artificial Intelligence makes it possible to have symbols with the value unknown or undefined. In data processing, it is the system developer, not the system, who determines all the relationships among the symbols. Knowledge Based Systems have the capability, depending upon the situation, to determine relationships between symbols which were not explicitly coded. A Knowledge Based System can do this because it has rules for manipulating relationships between symbols, which are explicitly declared.

The necessity to deal with relationships becomes obvious in comparing the notions of data, information and knowledge. Data, as used in data processing, is every kind of value which is accessible by a computer for processing. Information, as used in information processing, can be described as data that has been selected and organised for a particular purpose. Knowledge, as used in Artificial Intelligence, is information structured in such a way that brings out and exploits the relationships among the pieces of data or information. Methods of Artificial Intelligence use not only accessible data but also the knowledge embedded in the relationship between data. Knowledge processing is characterised by the following features:

- An inference engine utilises knowledge to find a solution for a given problem, e.g. path finding for a robot manipulator. This knowledge is specific to the problem domain and explicitly represented in the knowledge base.

- Declarative knowledge describes the possibilities of problem solving by the system, e.g. axioms and inference rules in Mathematics. The quality of the solution depends mostly on the quality of this knowledge.

- Meta-knowledge, which is meant here as knowledge about knowledge processing, controls the strategy for finding the problem solution. In a real time expert system, for instance, meta-knowledge would determine the knowledge sources which are helpful to obtain a coarse solution and then, if time is left, order a more detailed solution.

- In many problem domains, fuzzy knowledge is involved and has to be processed by the system. Facts are incomplete or not precisely known. Rules can be trusted only to some degree. Relationships between data and meaning are missing, e.g. between symptoms and diagnosis. Knowledge Based Systems are able to perform well in such domains. They possess rules such as how to proceed in ill-conditioned situations, how to rate hypotheses, and how to employ redundant knowledge sources.

The scope of research in Artificial Intelligence is the development of methods which enable the reproduction of such human capabilities as understanding, recognition and generation of natural language and visual information, planning and problem solving. The explanation of a solution path to the user is also supported by the methods developed. The industrial application of Artificial Intelligence methods will increase the variety of problem classes which can be solved automatically by computers. Diagnosis, monitoring, configuration, control, planning, advice, consultation and deduction are only some of the tasks which are already successfully carried out by Knowledge Based Systems.

3. Technological environment

In the last years, Artificial Intelligence has stayed in the focus of interest in scientific publications as well as in industrial publications [Schachter 85b]. The progress in the science of Artificial Intelligence triggered the publicity, with the driving force being the application of AI-techniques – resulting in Knowledged Based Systems – to industrial and commercial problems. Applications of AI are becoming practical and of commercial value due to several facts:

- Decision Support Systems are spread into broad use; they incorporate knowledge based technology together with a comfortable, user-friendly man–machine interface (e.g. spreadsheet programs).
- Computer hardware costs are decreasing, arriving at a level where it is economically reasonable to provide AI-based technology for broad practical applications.
- Complete high-level, graphic oriented workstations are available to become integrated in the process of system development and to provide the developer with new types of highly interactive tools.
- In a stripped-off version, those workstations will become cheap and powerful enough to serve as runtime systems for the end-user. They will be equipped with integrated communication facilities to be easily embedded in conventional environments.

4. Classes of knowledge based systems

Genuine domains for AI systems are:

- natural language understanding,
- perception,
- learning, and
- advanced problem-solving.

Natural language understanding systems analyse and generate written language. Two problems have to be solved. They have to interpret the input text to unravel its semantic contents. Furthermore, they have to build up a model of their dialogue partner, concerning his interests, his intentions, his main topic, and to judge his level of information. Such systems help casual and therefore unexperienced users, e.g. to interact with an information or reservation system, or to get acquainted with a new tool, e.g. a CAD-system, in a tutoring situation. The rigid dialogue of conventional systems can be overcome to make the computer a more acceptable communication partner.

Perceptive systems understand visual or acoustical and in some robotics applications even tactile information. They do this in forming a more and more precise internal model of external, sensed environment. This enables them to infer reasonable actions depending from the state of the model and the goals. Speech understanding and generation in combination with natural language understanding will lead to a fluent conversation with a computer. Image understanding systems found their place in industrial environments to automise manufacturing processes. They monitor the quality of workpieces, guide manipulators, or survey rooms or dangerous areas. Medical applications are still an important and motivating area for image understanding, e.g. the interpretation of Roentgen-images, scintigrams, ultrasound recordings, and computer- tomograms.

Learning is concerned with the acquisition of new knowledge. Of more interest than merely to acquire facts is the problem solving that is required to integrate into the system new knowledge that is presented to it. Another facet is to deduce new information when required facts have not been presented. The success of learning systems will improve the quality of Knowledge Based Systems, especially they will ease the still cumbersome process of knowledge acquisition and maintenance. However, it will last some time until these systems are available for a broad industrial application.

Advanced problem-solving is commercially used in planning and Expert Systems. **Planning** is the activity which decides what actions are to be taken in order to achieve a given goal. To manage complex technical process plans under time constraints is a challenge for Knowledge Based Systems. To describe an assembly process not by the teaching of coordinates but rather by the result to be achieved, specialised but flexible, intelligent planning systems are needed. They should not only optimize some criteria according to an evaluation function, but rather apply planning strategies to their model world to create complete and correct solutions. Computer Integrated Manufacturing is a main area where CAD, CAM and MRP techniques are combined by Knowledge Based Systems to build up the factory of the future.

Expert systems employ a knowledge base with facts and rules together with an inference engine to solve the problems of their users. They are interfaced to the external world to allow them a dialogue with a user or the controlling and monitoring of a technical process. An incorporated explanation facility improves the acceptance of expert-as well as planning-systems by users. This component explains which facts and rules contributed to the presented solution. A knowledge acquisition module is essential for the maintenance of the knowledge base. This component provides the facility to change, add, and remove rules or facts in an easy, interactive manner by the knowledge engineer or the expert. It helps them to evaluate the consequences of these modifications.

5. The life-cycle paradigm for the software development

The software development process can be well described by a life-cycle paradigm. The possible life-cycle models provide a uniform framework for problem solving within which reusable methodologies and tools can be developed. Two approaches are typical for the state of the art:

Waterfall models
requirements → design → implementation → maintenance

and

Operational models
executable specifications → transformations → efficient implementation.

Waterfall models are characterised by the software development proceeding through a number of stages. Each stage has documentary output that serves as the input to the next stage. Early stages specify a behavioural abstraction of **what** is computed. This abstraction is progressively refined into a formal implementation of **how** the behaviour can be realised. Maintenance and enhancement is performed on the implemented program. The life–cycle model used in PRADOS – 'Projektabwicklungs- und Dokumentationssystem der SCS' (Fig. 19) shows all previously mentioned aspects, but allows continuous refinement by iterations. As a result, the life-cycle model moves in the direction of an operational model.

In **operational models** software development proceeds from an executable problem-oriented specification **(rapid prototype)** through a sequence of transformations to a more efficient implementation-oriented realisation. Early stages are independent of computational resources. Operational abstraction

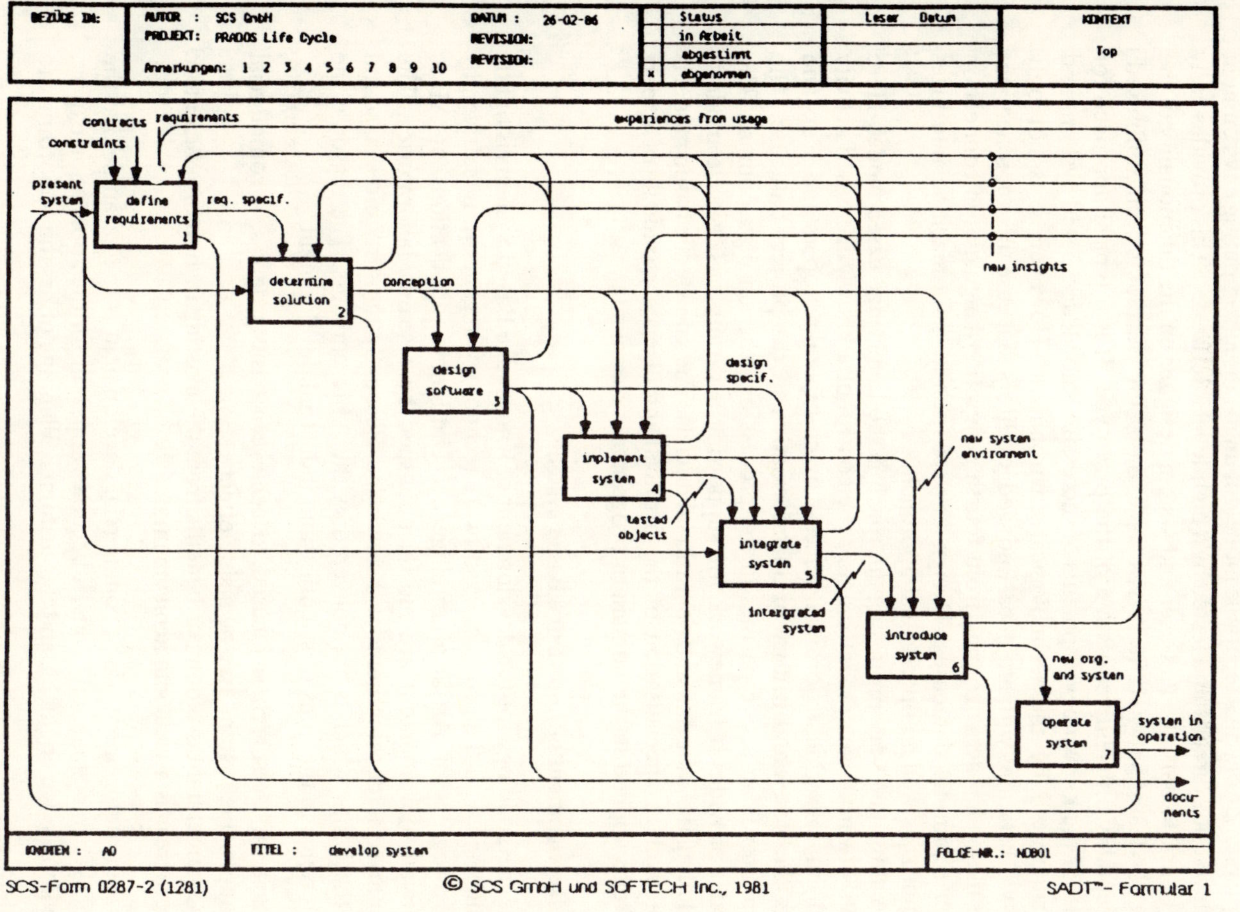
BEZÜGE IM:
AUTOR : SCS GmbH
PROJEKT: PRADOS Life Cycle
Anmerkungen: 1 2 3 4 5 6 7 8 9 10
DATUM : 26-02-86
REVISION:
REVISION:
Status
in Arbeit
abgestimmt
abgenommen
Leser Datum
KONTEXT
Top
contracts
requirements
constraints
present system
define requirements 1
req. specif.
determine solution 2
conception
design software 3
design specif.
implement system 4
tested objects
integrate system 5
intergrated system
introduce system 6
new system environment
operate system 7
new org. and system
system in operation
documents
experiences from usage
new insights
KNOTEN : A0
TITEL : develop system
FOLGE-NR.: NO801
SCS-Form 0287-2 (1281)
© SCS GmbH und SOFTECH Inc., 1981
SADT™- Formular 1

Fig. 19

determines a process-oriented paradigm for problem solving. Instead of verbal requirements and a design that determines the modular structure of the implementation, an executable model for an idealised computing environment is developed directly and tested by a high-level interpreter. This provides early feedback to both the end-user and system designer on the functionality of the intended system, and it serves as a basis for developing an optimised, acceptably efficient realisation of the prototype. It is a typical approach for most of the known developments of successful Knowledge Based Systems and is supported by the technological environment. However, it is not only AI-workstations with graphical knowledge editors that ease this approach, but the major impact is coming from fourth generation languages in connection with relational databases. Design of user-friendly man-machine interfaces requires as well an operational approach.

A third and very promising approach is the **knowledge based model**. Here the software development is under the control of a knowledge based activity coordinator, which coordinates access by multiple developers and users and logs the states and history of all information in the project database. The computer is an active partner in the process or program development. The model provides a framework for supporting a variety of life-cycle models, including the waterfall and operational models. The knowledge based model is effectively a specialisation of the artificial intelligence paradigm of expert systems applied to the domain of program development.

6. Development of knowledge based systems

The methodology for the development of Knowledge Based Systems is tightly connected with the progress of software technology in general. It is driven by the progress of Artificial Intelligence. Until now, applications of AI to software development are mostly published by research laboratories. The commercial development of Knowledge Based Systems requires a sound methodology which is not delivered by the AI researchers. The reason is the difference in the aims of scientific and industrial development:

- Scientific research is focused to achieve a basic progress, a commercial applicable system may be a side effect.
- Commercial development benefits from the existing technology. Scientific progress is a welcome achievment.

A main requirement of the commercial development is to guarantee that the process can be planned. Thus, it becomes apparent for the application expert, the knowledge engineer, and the customer. Planning offers the opportunity to

- calculate and monitor costs,
- to monitor the progress of the project with respect to time, resources, goals and success, and
- to supervise the quality of the final product.

In an industrial environment, not all members of the development team are equally experienced and qualified in all aspects of KBS development. Therefore, a methodology has to divide the development into phases and define representations for recording the results of each phase. This representation should allow a clearer understanding of the relationship between the function of the system, the structure of the knowledge in the domain, and the design structure as well as the architecture of the implemented system. Unfortunately a pure waterfall model is not taking into account the experience in building Knowledge Based System nor the intrinsic complexity of such a system which cannot be specified only by an informal behavioural abstraction. Prototyping is typical for the specification of knowledge based systems and suitable for their incremental development. But an approach purely based on an operational model might be dangerous for the development in large industrial teams.

Driven by these insights the project 1098 [KBSM] of the ESPRIT program was setup. This project addresses directly the objectives given under Area 3.5.1 of the ESPRIT workplan (A life cycle model of AIP systems), namely to:

- identify an adequate range of effective methods and techniques of AIP system production and maintenance, giving reliable, secure and tolerant systems;
- provide tool support for those methods and techniques, assuming several layers of representation with incremental verification and validation.

The purpose of the project is to assist the transfer of Knowledge Based Systems technology into commercial use by providing methodological guidance for the development process. To achieve this issue a structuring of the development process is necessary by partitioning it into discrete, well-defined phases, each of those being supported by related techniques and tools. The result is a waterfall model of the development of KBS (Fig. 20), **activity** being the central concept. Activities are composed into phases. Each phase has a well-defined result as output, which is a prerequisite for the beginning of the following phase(s). These results also represent the actual status of the development process, so the documentation grows as the software development project proceeds. Activities are structured by defining new subactivities. A hierarchy of activities is given by this structure (Fig. 21, Fig. 22). The structural similarity between the PRADOS model and the KBS lifecycle model is obvious and is underlined by using the same formal representation: SADT (Structured Analysis and Design Technique). From both life-cycle models the same guidance for the developer can be derived:

- what has to be done in which phase
- where dependencies exist between activities
- which prerequisites must be fulfilled in order to start a certain activity
- which results are provided by an activity or phase

– which feedbacks may appear in later phases and what consequences they have.

The representation of the models is permissive with regard to iterations and in this respect not typical for waterfall models. Specifically, the KBS–LCM moves towards an operational model giving credit to an incremental specification through prototyping (Fig. 20, Fig. 21). Basic ideas of operational models like early feedback to both the end-user and system designer on the functionality of the intended system are taken into account. Prototyping may accompany the first three phases (determine requirements, acquire knowledge, design system) leading after several iterations from an executable model for an idealized computing system environment to a more efficient implementation-oriented realisation.

The knowledge acquisition phase is structured by modelling the three stages [BREUKER 85a]: orientation, problem identification and problem analysis, each of those iterating over activities of elicitation and analysis of knowledge. Knowledge analysis provides an interpretative frame work which is used in further elicitation and analysis. This approach is necessary in the context of an activity centred model.

An operational model would represent the lifecycle of a KBS as a transformation process through five levels of representation of knowledge [Brachman 79, Newell 80]. These levels are: linguistic, conceptual, epistemological, logical, and implementational. A proper mapping of these levels to the three mentioned stages in which analysis occurs is not obviously possible.

The best guidance to the developers combining the relevant aspects of a waterfall and an operations model is given by a knowledge based model for the software development. As previously mentioned a knowledge based model is a specialization of the expert systems paradigm applied to the domain of software development; the core of such a life-cycle model is the implementation of the methodology into a Knowledge Based System. The Knowledge Acquisition and Documentation System (**KADS**) [Breuker & Wielinga 85] is the methodology and support environment developed in ESPRIT project 1098 (KBSM].

KADS is also a knowledge engineering toolbox for various aspects of the development process:

Formalisation

An activity oriented lifecycle is the development framework. The results of each activity is described.

Guidance

The toolbox can be used as a source of instructions for acting, including advice in the interpretation of acquired information.

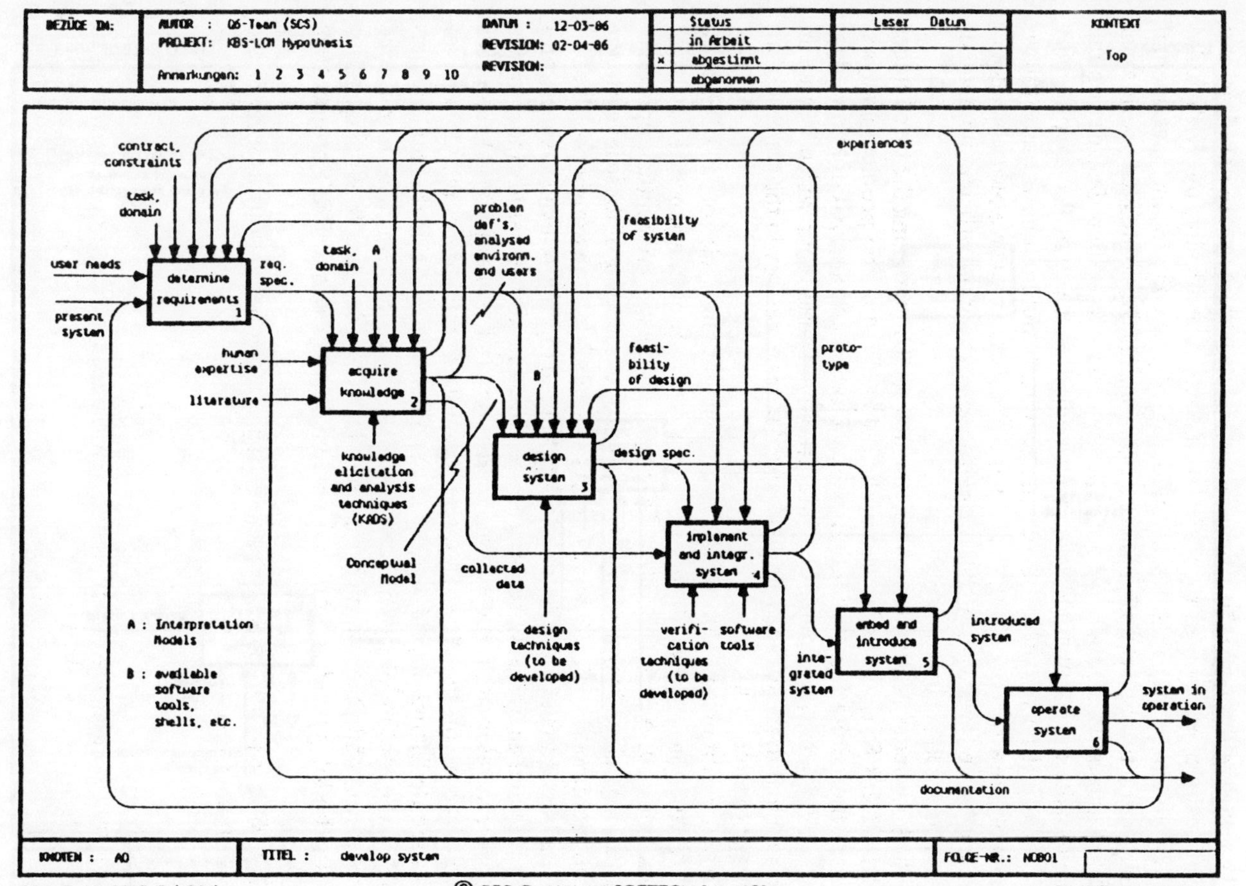

SCS-Form 0287-2 (1281) © SCS GmbH und SOFTECH Inc., 1981 SADT™- Formular 1

Fig. 20

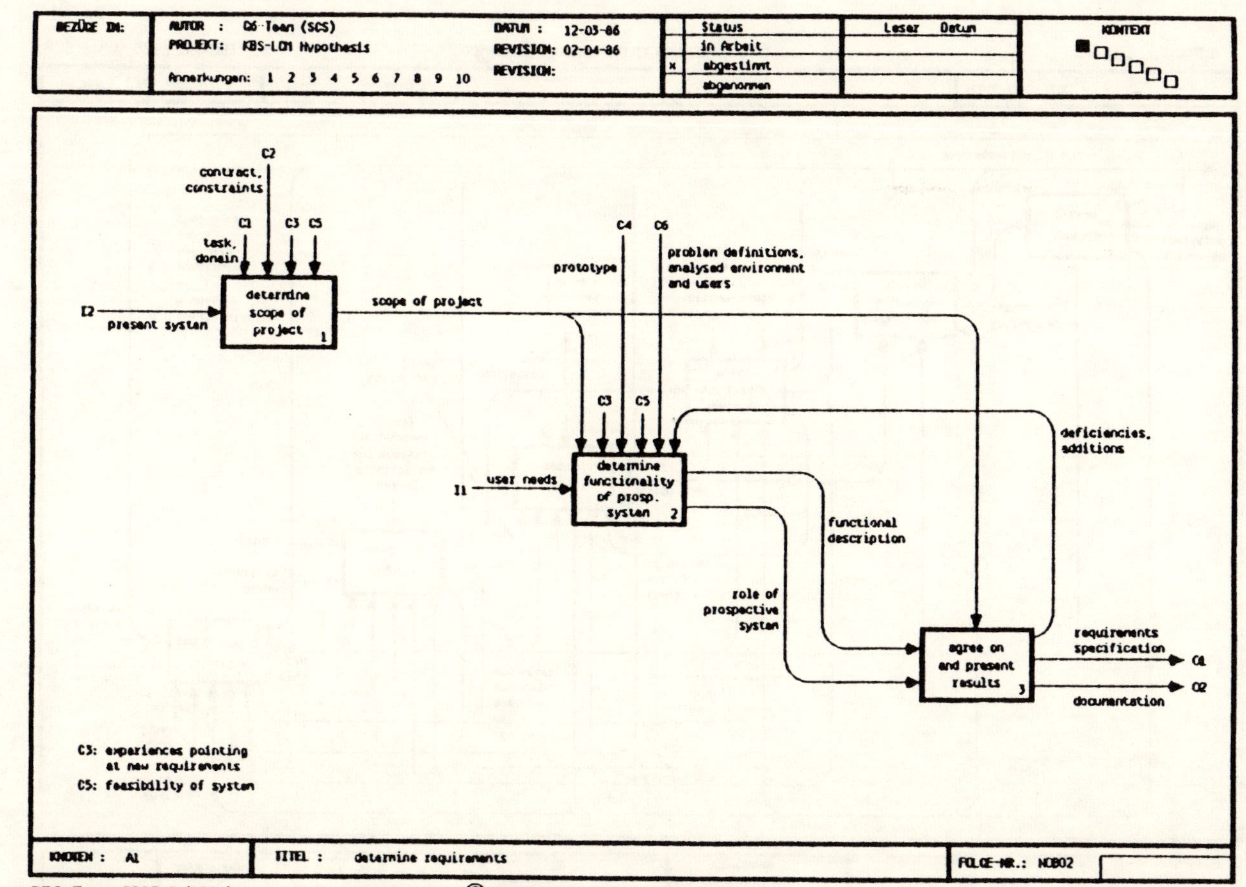
BEZÜGE IN:
AUTOR : Q6-Team (SCS)
PROJEKT: KBS-LCM Hypothesis
Anmerkungen: 1 2 3 4 5 6 7 8 9 10
DATUM : 12-03-86
REVISION: 02-04-86
REVISION:
Status
in Arbeit
x abgestimmt
abgenommen
Leser Datum
KONTEXT
C2
contract, constraints
C1
task, domain
C3 C5
I2
present system
determine scope of project 1
scope of project
C4 C6
prototype
problem definitions, analysed environment and users
C3 C5
I1
user needs
determine functionality of prosp. system 2
deficiencies, additions
functional description
role of prospective system
agree on and present results 3
requirements specification
O1
documentation
O2
C3: experiences pointing at new requirements
C5: feasibility of system
KNOTEN : A1
TITEL : determine requirements
FOLGE-NR.: NO802
SCS-Form 0287-2 (1281)
© SCS GmbH und SOFTECH Inc., 1981
SADT™- Formular 1

Fig. 21

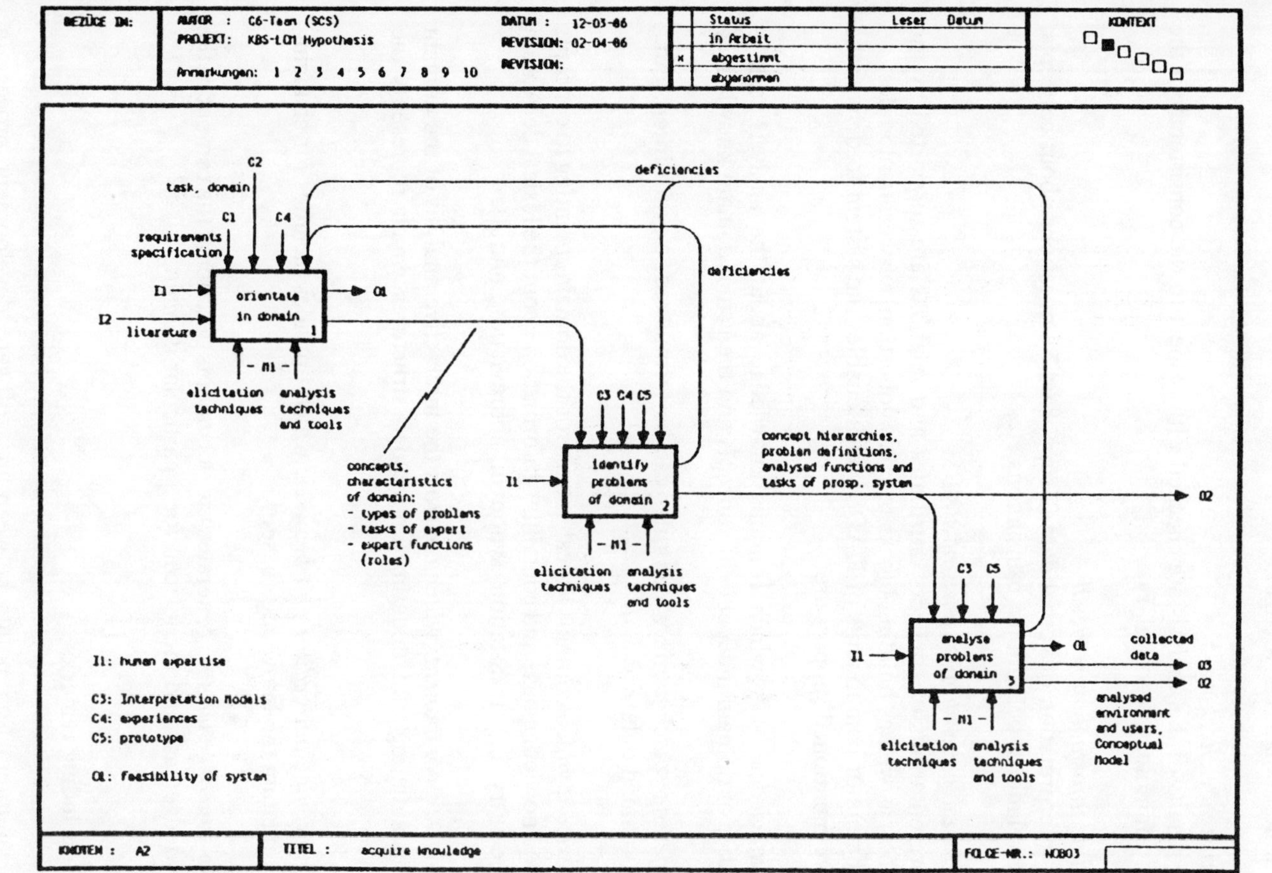
BEZÜGE ZU:
AUTOR : C6-Team (SCS)
PROJEKT: KBS-LCM Hypothesis
Anmerkungen: 1 2 3 4 5 6 7 8 9 10
DATUM : 12-03-86
REVISION: 02-04-86
REVISION:
Status
in Arbeit
x abgestimmt
abgenommen
Leser Datum
KONTEXT
C2
task, domain
C1
C4
requirements specification
I1
I2
literature
orientate in domain 1
O1
- M1 -
elicitation techniques
analysis techniques and tools
deficiencies
deficiencies
concepts, characteristics of domain:
- types of problems
- tasks of expert
- expert functions (roles)
C3 C4 C5
I1
identify problems of domain 2
- M1 -
elicitation techniques
analysis techniques and tools
concept hierarchies, problem definitions, analysed functions and tasks of prosp. system
O2
C3 C5
I1
analyse problems of domain 3
O1
collected data
O3
O2
analysed environment and users, Conceptual Model
- M1 -
elicitation techniques
analysis techniques and tools
I1: human expertise
C3: Interpretation Models
C4: experiences
C5: prototype
O1: feasibility of system
KNOTEN : A2
TITEL : acquire knowledge
FOLGE-NR.: NO803
SCS-Form 0287-2 (1281)
© SCS GmbH und SOFTECH Inc., 1981
SADT™- Formular 1

Fig. 22

Documentation
The structure for the documentation is given, to ease the maintenance of the acquired domain knowledge.

Consistency checking
The results of the analysis are checked with respect to the consistency of the user teminology (vocabulary).

On-line information retrieval
The interpretation of verbal data is supported by questions about concepts, their relations, definitions, translations etc.

Automatic generation of parts of a KBS
Representations of concepts, attributes and relations are supported by the toolbox as well as building-blocks to create question–answering modules.

The major components of KADS which enables the system to deliver the above mentioned support are:

Knowledge base: which contains knowledge about the analysis task, a number of domain independent concepts and the actual domain knowledge;

Inference machine: in which rules can be formulated, e.g. for interpreting the obtained data;

Lexicon: which contains lexical entries for the domain, pointing to concepts in the knowledge base and containing information about the lexical derivation of the entry (e.g. translation, synonym, abbreviation, plural);

Analysis component: which controls the interactive analysis of the domain by instantiating and updating knowledge structures which represents the analysis task;

Knowledge base editor and browser: which provides facilities for changing and updating the knowledge base;

Document generator: generates the documents describing the content in the knowledge base and the lexicon for a particular domain.

7. Concluding remarks

Several interaction points between the software development process in general and the Artificial Intelligence paradigm were presented. Knowledge Based Systems become more and more attractive for solving various automation areas independently of the application domain. Artificial Intelligence is an enabling technology, and one of the application domains is software production. There are various approaches to use this technology [Broy 85], KADS methodology is a development into the future towards a completely supported knowledge based life-cycle model for the development of Knowledge Based Systems. A similar approach can be as well taken for the development of more conventional information systems, the expertise necessary to build up an expert system environment for software engineering is available.

Acknowledgements

I would like to take the opportunity to acknowledge the originality of the ideas concerning KADS to Dr. Joost Breuker and Dr. Bob Wielinga of the University of Amsterdam and to express my gratitude to Simon Hayward from STC for transforming ideas into a hopefully successful project.

For many fruitful discussions I would like to thank all my coworkers in the ESPRIT project 1098, especially those involved in the assessment of life cycle models. The interest in software engineering problems amongst my colleagues at SCS encouraged this presentation.

Last but not least, I would like to thank my husband, Prof. Bernd Radig of the Technische Universität München for not giving up to criticising my work.

References

Bauer 85 BAUER, F. L., 'Geleitwort': in BRAUER, W., RADIG, B. (eds.), 'Wissensbasierte Systeme', GI-Kongreß 1985, Informatik-Fachberichte 112, Springer-Verlag Berlin Heidelberg 1985

Brachman 79 BRACHMAN, R. J., 'On the epistomological status of semantic networks': in FINDLER, N. V. (ed.), 'Associative networks', Academic Press, New York, 1979

Breuker 85 BREUKER, J., WIELINGA, B. 'Use of Models in the Interpretation of Verbal Data': in KIDD, A. (ed.) 'Knowledge elicitation for expert sytems: a practical handbook', Plenum Press, New York

Breuker & Wielinga 85 BREUKER, J., WIELINGA, B. 'Structured Knowledge Acquisition for Expert Systems', University of Amsterdam, VF Memo 40, SWI/University of Amsterdam, and Proceedings of the Fifth International Workshop on Expert Systems and their Applications, Avignon (France), 13–15 May 1985

Broy 85 BROY, M. 'Rechnergestützte Systeme für den Programmentwurf: in BRAUER, W., RADIG, B. (eds), 'Wissensbasierte Systeme', GI-Kongreß 1985, Informatik Fachberichte 112, Springer-Verlag Berlin Heidelberg 1985

Hart 86 HART, P. E., BARZILAY, A. and DUDA, O., 'Qualitative Reasoning for Financial Assessments', AI magazine, Vol. 7, No. 1, Spring 1986

Koen 85 KOEN, B. V., 'Toward a Definition of the Engineering Method', The Bent, Spring 1985, pp. 28–33

KBSM 'Technical Annex ESPRIT Project 1098', STC plc, Scicon Ltd, SCS Organisationsberatung und Informationstechnik GmbH, Cap Sogeti Innovation, The University of Amsterdam, and the KBSC (South Bank Polytechnic, London)

Newell 80 NEWELL, A, 'Physical symbol systems', Cognitive science, 1980, 4, pp. 135–183

Schachter 85a SCHACHTER-RADIG, M.-J., 'Wissenserwerb und -formaliserung für den kommerziellen Einsatz Wissensbasierter Systeme': in BRAUER, W., RADIG B. (eds), 'Wissensbasierte Systeme', GI-Kongreß 1985, Informatik-Fachberichte 112, Springer-Verlag Berlin Heidelberg 1985

Schachter 85b SCHACHTER-RADIG, M.-J., 'Knowledge based systems – new phases in the project development', Proceedings of the CAMP '85, Berlin, 1985

Scown 85 SCOWN, S. J. 'The Artificial Intelligence Experience: An Introduction', 1985, Digital Equipment Corporation

Wegner 84 WEGNER, P. 'Capital Intensive Software Technology', in IEEE Software, Vol. 1, No. 3, July 1984

Automating the transformational development of software

Guido Persch

GMD Forschungsstelle, an der Universität Karlsruhe, Haid-und-Neu-Strasse 7, D-7500 Karlsruhe, West Germany

Keywords: Software Engineering, Transformational Development of Software, Requirements Definitions, Algebraic Specification, Prototyping, Expert Systems, Ada.

Abstract: The transition from one document within a software development to the next can be viewed as transformation of the underlaying languages. Depending on the determinism and completeness of the documents this transformation can be automised or supported by a computer. In this paper we define a model of a software development environment based on an expert system for transformations. The transformation system behaves similar to a Prolog system. We discuss the structure of the system and the tasks of the different components. For the phases requirements, engineering specification, prototyping and implementation we show examples of the system within such applications.

1. Introduction

Software development is normally divided in phases like requirements definition, design, implementation and maintenance to get separated working units. Such a phase model can be tailored to special needs of a project. The products of each phase are documents like requirements specification, system specification, and programs. These documents are formulated in languages which are designed for the phases, like stimulus-response-nets for requirements specification, algebraic specifications for system specification, Ada for implementation. The process of software development has to start from an idea of a software system and to generate the documents for the different phases [LEH 83]. In the ideal case the documents of subsequent phases are transformations (in a very general sense) of the first documents in order to get more detailed, deterministic and executable descriptions of the software system [BAU 81]. This transformational approach of software development has only been used informally in the past. New developments [BRO 83], [BAL 82] try to apply this approach with explicit support and sometimes in a rigorous way. Therefore we have to provide three prerequisites:

- Formal definitions of languages to express the documents of the transformations
- Definition of a transformation calculus which allows the formulation of transformation rules and derivations
- Development of support systems to automise transformations

The CIP-system [PAR 83] is a transformational system based on an algebraic wide-spectrum language. From a theoretical point the CIP-System has been a first step to provide the basis for the transformational approach. The implemented system however is rather restricted. It requires deterministic transformation rules and does not support automatic rule selection. Hence the programmer has to guide the transformations on a very low level.

The GIST-System [BAL 82], [FIC 85] is a transformational development system for a special specification language GIST. GIST is a very high level operational language with non-deterministic constructs (demons). Target language of the transformations is a special implementation language. The system is tailored to these languages. It provides a high degree of automation in respect to rule selection but it is not based on a formal calculus.

An essential aspect of improved software development is the introduction of formal descriptions in the first phases (requirement specifications). This has been delayed because such documents could not be further processed. A formal requirements specification is only useful if it can be validated and if it is the basis for further developments. With transformation systems it is possible to transform formal requirements specifications into system designs and finally implementations. They also provide means to formulate specification languages which are tailored to the application field. Such languages represent implicit knowledge about the applications which is then expressed explicitly with the help of the transformation system in the system design.

2. Model of an environment

The general structure of a transformational support environment has three levels, the user interface, the tools, and the data base. The user interface has to provide the means for user–machine communication especially window management and graphical support. The tools consist of conventional tools like editors, analysers, compilers and the transformation system as essential part. The transformation system is the supervising part of the tool level.

Besides the pure language transformation rules it contains also rules for the activation of other tools. An example is the control of the compilers to get a configurated system out of parametrised modules where the rules include the configuration of the sources and arguments for the compiler. The data base manages the documents which are generated during the development process, status information, and also the rule sets which are needed to process the documents. In order to ease the application of tools (especially of the transformation system) the documents are stored in structural form and not as texts. Therefore an entity/relationship database system [CHEN 76] is required which allows to store the documents in a structured way and to keep the relationships between parts of the documents. Such a database has influence on the tools like editors which then are also structure-oriented (i.e. syntax-directed editors). A further advantage of such a database is the possibility to handle graphical languages (like SADT, stimulus-response-nets). The transformation rules stored in the database are separated into rule bases. A rule base can represent the knowledge for an application field (e.g. transformation of stimulus-response-nets into Ada programs) or it contains actually applied rules to record the development history.

Software development is done in three working steps: editing, analysing and transforming. Editing is used for initial document generation and document modification. Analysing checks the documents for consistency and completeness (static analysers) or validates them by execution or simulation (dynamic analysers like interpreters, debuggers). Transforming maps the documents from one phase to another or improves documents within one phase. Such transformations can range from highly interactive to full automation. If no strategy is clear (especially for new application fields) the user guides the transformation or interactively introduces new rules. If a development is done a second time with nearly the same documents or the choices are rather restricted the system can itself run the transformation maybe in a trial-and-error way.

The advantages of this development process are:

Correctness: If later documents are generated by transformations with verified rules from an initial document, then these documents are correct in respect to the initial document (correctness by construction). However, the validation of the initial document and the verification of the rules are not trivial but can be done on a much higher level than the validation of each development product.

Productivity: The automation of simple transformation tasks which require most of the time at least during the implementation phase disburdens the developer and guarantees consistent performance. (E.g. substitute all calls of a subprogram by its body, propagate its actual parameters and simplify.)

Prototyping: The transformation of the first documents of a project into very high level languages allows rapid prototyping in early phases. Hence errors and misunderstandings could be avoided very early. Prototyping done by hand is normally too expensive to apply in every project.

Reusability: After keeping the transformation rules which reflect the development of some class of applications they can reused in further projects (development reuse). It is essential, that such rules are classified in proper rule bases to reduce the complexity of rule catalogues. Libraries of fixed program modules are not flexible enough to guarantee reusability [YEH83]. The adaptations which have to be done can't be expressed in parameterised modules but they can be expressed by parameterised developments.

Maintainability: The transformation system allows to do the changes in the specification and to replay the development. Furthermore the formalisation of the development history by the sequence of transformations allows to backtrack to decisions and to change them. Afterwards the new implementation can be done by replaying the development. If the replay succeeds the changes are compatible with the remaining system parts, otherwise the concerned parts can be derived and changed as well.

Disadvantages of a transformational software development environment could be:

Rule base size: The number of rules grows to sizes where reasonable machine handling is not possible. This should be avoided by classification of rules into rule bases which are tailored to applications and compilation of rule bases which are completed.

Efficiency: The implementations generated by an (automatic) system are less efficient than manual implementations. The reasons are missing formalizations for implementation techniques. On one side this can be accepted because there are other advantages and on the other side such formalizations should grow by using transformation systems.

Readability: Documents generated by an automatic system are hard to read for the human reader and very difficult to maintain. If such documents are processed by machine it does not matter. The actions performed by the transformation system can be so complex that it is difficult for the user to follow. In such a case an explanation subsystem should be there.

3. An expert system for transformations

A transformation system to be used in a transformational development environment is best constructed like an expert system. It consists of three subsystems, the rules, the inference machine, and the explanator.

3.1. The rules

In general transformation rules have the form (cf. [BAU 81]):

Input Pattern —Condition—> Output Pattern

with the semantics, that a structure which matches the input pattern is transformed to the output pattern if the condition holds. The pattern and the condition share some variables which are instantiated during the match of the input structure. This general scheme is extended for our system by the following:

- *Rule names* are assigned to groups of rules. This introduces some kind of modularisation for rules and allows to reference rules.
- Rule names can have *parameters* to fomulate meta rules which control the application of rules. By this mechanism rules can also be combined to larger rule sets.
- The parameters in rule names also allow to transport information from a global context to the place of application where it can be used in conditions and other applications.
- The patterns allow variables for function symbols and for parameter lists of functions.
- Output patterns can contain *rule applications* with the semantics of substituting them by one of the results of the invoked transformation. By such means a transformation rule can call subrules which are then composed to reach the required transformation.

The documents to be processed are represented as term data structure which corresponds directly to a abstract syntax representation of programs in compilers. This abstract syntax representation has to be provided by the database. The patterns in the transformation rules are formed in correspondence to it.

In our Prolog implementation of the transformation system, rules have the form:

rule rule_name : input_pattern => output_pattern **cond** condition

Within the output_pattern and the conditions rule invocations are written as *rule_name* : *input*, patterns for functionals are written as *variable#arguments*. Rule names are terms with variables. Variables are marked by their beginning big letter. Conditions are conjunctions, disjunctions and negations of predicates.

Example: Rules for Expression Simplification

rule *simple_exp:* Exp => *inner(simple):* Exp.

rule *simple*: plus(X,0) => X.
rules *simple*: times(X,1) => X.

rule *simple*: OP#[C1,C2] => C3
  cond constant(C1) **and** constant(C2) **and** eval(OP,C1,C2)=C3.
rule *inner(R)*: F#[X] => R: F#[*inner(R)*: X].
rule *inner(R)*: F[X,Y] => R: F#[*inner(R)*: X, *inner(R)*: Y].
rule *inner(R)*: X => X **cond** atom(X).

The example consists of one top rule *simple_exp* which has two subrule bases *simple* and *inner*. The *simple* rules describe the different simplifications, whereas *inner* describes the strategy to apply these simplifications. Hence, *inner* can be seen as meta rule.

The third *simple* rule describes that any dyadic expression with constant arguments can be substituted by its result. Therefore we use a pattern for any dyadic expression with the operator OP. The *inner* rules have as parameter the name *R* of a rule base which is to be applied first to the innermost subexpression and then to enclosing expressions. Additionally we see how the output of *inner* is composed of the results of the invocations of *inner* to the subexpressions and the invocation of *R* to this composition.

3.2. Inference Machine

The inference machine of an expert system has the task to apply the set of rules to the data. Hence it has to solve the following problems:

Pattern Matching: Pattern matching is done by an extension of unification for first order logic (e.g. an extension of pattern matching in Prolog). The extension handles variables for function symbols and for lists of parameters.

Condition Evaluation: If the input pattern matches the data the conditions have to be evaluated with the bindings done for the variables. Evaluation of conditions is handled like transformations to the boolean values. If they succeed the logical connectives are evaluated. Some of the conditions may require more sophisticated evaluation for example by a theorem prover. Additionally some of the conditions should be handled interactively, i.e. the programmer is asked for a decision.

Order of evaluation: Most of the order evaluation is already formulated by meta rules, which describe the rules to be applied to some data. However, there can be several rules (with the same name) which can be applied. In this case a trial-and-error evaluation proceeds, i.e. the rules are applied in some order until one succeeds.

Blind alleys: If no more rule is applicable for a rule invocation, this can result from a wrong rule applied earlier. In this case the inference machine backtracks to a position where another rule can be applied. A transformation only succeeds if all rule invocations (which can invoke further rules) are successful. Hence the transformation is goal-oriented. It backtracks from blind alleys automatically.

Termination: A transformation can invoke rules forever especially if there are commutativity and associativity rules. With the help of meta rules such rules can be reformulated in a safe way, e.g. by applying commutativity only once to a term.

Example: Evaluation of expression simplification

simple_exp: times(a, 1)
=> *inner(simple)*: times(a, 1)
=> *simple*: times # [*inner(simple)*: a, *inner(simple)*: 1]
=> *simple*: times(a, 1)
=> a

In the current implementation of the transformation system, the transformation rules are mapped to Prolog clauses and a standard Prolog interpreter builds the inference machine. Because of the efficient implementation of Prolog interpreters this implementation is useful. In the future a rule compiler (which is a set of transformation rules for transformation rules) will be added which translates complete rule bases to programs which are called at the rule invocation.

3.3. Explanator

The explanator has the task to inform the programmer of the applied transformations. This is separated into two cases:

Success: If a transformation succeeds all the rules which have been applied are recorded in a history list. This list can be extended by the alternatives which are still possible. The list can be prepared for the user by suppression of the non-interesting rules.

Failure: If a rule invocation fails or the system cannot prove a condition (or its contrary), the status has to be displayed in a way that the programmer can interact.

In the current Prolog implementation of the transformation system all applied rules are recorded and can be given during the transformation as a trace or afterwards as a history list. If during the evaluation of conditions undefined predicates are found, the system gives status information to the user like rule names and current part of the document and requires interactive use of the programmer.

4. Applications

In this chapter we show applications of the transformation system for the development of process control software. We assume a life cycle consisting of requirements definition, specification, prototypying and implementation.

4.1. The Development Process

Stimulus-response-nets (SR-nets) are an established method of requirements definitions for process control systems [ALF 80]. We use the language LARS [EPP 83a] for the requirements specification. LARS is a modified variant of RSL which is used in SREM. For LARS exist a graphical editor [EPP 83b] and an analyser, which allow a user-friendly preparation of requirements. LARS specifications consist of three parts: a *guard*, which controls all the activities of the system, the *SR-nets*, which describe what action (response) is to be performed for a given event (stimulus), and the *interfaces*, which describe the interfaces to the technical process to be controlled. Data and its operations are described in natural language in LARS. Non-functional requirements like time constraint are already fixed in the LARS document.

The formalisation of the data and its operations has to be done in the system specification phase by algebraic specifications [GOG 84]. Data is described by abstract types which consist of a signature and axioms in the form of conditional equations. The abstract data types include exceptions which can be raised in error situations and which can be handled by other operations. The first complete formal system specification is a mixture of LARS and algebraic specification. This document is the starting point for the transformation during the prototyping phase and the implementation phase.

During the prototyping phase the specification is transformed into an Ada program. Prototyping is applied very early to validate the system specification. The use of Ada is prototyping language and implementation has the advantage of having only one language in both phases and allows to reuse existing software within prototypes. The goal of the prototyping phase is an Ada program, which has the functional behaviour of the final system and which already handles the interfaces. It can processed by an Ada compiler and Ada related tools [PER 85].

The implementation phase concerns mostly with efficiency, user interfaces and embedding in the technical process. The implementation can have a different system structure than the prototype. It has to observe especially all non-functional requirements.

4.2 Transformation

We describe the general proceeding in the transformation and show some examples of transformation steps.

The prototype has to reflect the four parts of the specification. The guard as central control of the system is transformed to an Ada task which waits for events from the interfaces, activates the SR-nets, and outputs data to the interfaces. The interfaces become entries of the guard task. The SR-nets which are parallel activities of the guard are transformed into tasks. Such a SR-net can be active several times in parallel. It can have local data which should be kept between several calls. Hence we have three possibilities for the transformation of SR-nets:

(1) a task object with permanent local data
(1) a task type where each SR-net activation generates a new task object with temporary local data
(1) a task type within a package where each SR-net activation generates a new task object with the global data in the package.

The decision for one of the alternatives requires the knowledge of the existence of permanent data and parallel activations. The first can be derived by data flow analysis from the specification, the second one is requested from the programmer.

Example: Rules for the Transformation of SR-Nets

```
rule net(GUARD): net(ID,GLOBALS,LOCALS,ALPHAS)   % case (1)
          =>  [task(ID, entry('start'))
             , task_body(ID,
                         decls: LOCALS,
                         net_body(GUARD): ALPHAS)]
          cond sequential_activation(GUARD,ID).

rule net(GUARD): net(ID,GLOBALS,LOCALS,ALPHAS)   % case (2)
          =>  [task_type(ID, entry('start'))
             , task_body(ID,
                         decls: LOCALS,
                         net_body(GUARD): ALPHAS)]
          cond not permanent_data(ALPHAS).

rule net(GUARD): net(ID,GLOBALS,LOCALS,ALPHAS)   % case (3)
          => [package(ID),
                 [task_type(ID), entry('start'))
                 [decls: LOCALS])
             , package_body(ID,
                         task_body(ID, . . .))]
          cond permanent_data(ALPHAS).
```

The abstract data types are mapped into Ada (generic) packages. The signature can be transformed one-to-one into a package declaration. The axioms are mapped to function bodies which operate on the term algebra for the abstract data type. If we separate the operations into constructors and defined operations, the constructors build the term data structure and the defined operations manipulate this data structure as defined by the axioms [PER 84]. By this technique the prototype implementation of the data types is automised.

In the implementation phase further transformations are applied to improve the abstract data type implementations. First rule bases for special data type implementations are evaluated. For example if there exist rule bases to implement sets by linked lists, by hash tables or by bit strings under certain conditions the system tries to evaluate which rule base is applicable for this specification. Further improvements can be done by evaluating whether there exists only one object of a type at a time. In this case the data is made local to the package and the functions are simplified by deleting the type from the parameter list.

Further transformations try to integrate several parts of the system into one module with the help of global analysis. We describe two techniques, the *combine* and the *pipe* technique.

combine [DAR 81] expands functions at their calling place and tries to combine them to a new function which avoids temporary data structures. A typical example is a function f1, which produces a sequence, and a function f2 which consumes this sequence. The combined function produces only one element and consumes it directly.

The *pipe* technique is an extension of the UNIX pipe mechanism. If some task t1 produces a sequence of elements which is consumed in the same order by some other task t2, then with the help of a buffer task tb, the construction of the sequence can be avoided. If t1 has produced an element it is given to tb, and if t2 will accept the next element tb gives it to t2. The evaluation of the applicability of these transformations is done on the specification level because the control and data flow analysis is much easier to handle there.

5. Conclusions

We showed a model of a transformational software development environment which can be instantiated for different application fields. The principal activities in the environment are editing, analysing and transforming. The basis of the environment is a transformation system which is a rule based expert system. This system has to apply given sets of rules in order to map documents from one phase to another. It has to cover the whole range from automatic to interactive transformation. Because of its flexibility such a system has a great advantage compared to program generators and compilers. It can be extended by the user's needs and tailored to his applications.

In the future it is necessary to develop an increasing set of rule bases which reflect the programmers knowledge. If such sets can be handled in a practical way, more and more tasks of the program development can be automised with the impact of improved productivity and correctness in software development.

References

ALF 80 M. ALFORD: SREM at the Age of Four, Proc. 4th COMPSAC, Chicago, IEEE 1980

BAL 82 R. BALZER, N. GOLDMAN, D. WILE: Operational Specification as the basis for rapid prototyping, Proc. ACM SIGSOFT Workship on Rapid Prototyping, Columbia 1982

BAU 81 F. L. BAUER, K. WOESSNER: Algorithm Language and Program Development, Springer Verlag 1981

BRO 83 M. BROY: Algebraic Methods for Program Construction – The Project CIP, Proc. Workship Program Transformations, Springer Verlag 1984

CHEN 76 P. CHEN: The Entity/Relationship Approach to Data Modelling

DAR 81 J. DARLINGTON, R. M. BURSTALL: A System which Automatically Improves Programs, Proc. IFIP Congres 1981, North Holland

EPP 83a W. EPPLE, M. HAGEMANN, M. KLUMP, G. KOCH: SARS – System zur anwendungsorientierten Anforderungsspezifikation, University of Karlsruhe, Inst. f. Informatik III, Report 1983

EPP 83b W. K. EPPLE, M. HAGEMANN, M. KLUMP, U. REMBOLD: The Use of Graphic aids for Requirements Specification of Process Control Software, Proc. 7th COMPSAC, IEEE 1983

FIC 83 S. F. FICKAS: Automating the Transformational Development of Software, USC-ISI/RR-83-108, Ph.D. Dissertation, Marina del Rey, 1983

GOG 84 J. A. GOGUEN: Parameterized Programming, IEEE Software Engineering September 1984, 5/10

LEH 83 M. M. LEHMAN, V. STENNING, W. M. TURSKI: Another Look at Software Design Methodology, Imperial College London, Report DoC 83/13

PAR 83 H. PARTSCH: The CIP Transformation System, Proc. Workshop Program Transformations, Springer Verlag F8-1984

PER 84 G. PERSCH: Designing and Specifying Ada Software, Proc. IFIP WG 2.4, Canterbury, North Holland 1984

PER 85 G. PERSCH et al.: Karlsruhe Ada System – User Manual, Gesellschaft für Mathematik und Datenverarbeitung mbH, Report 1985

YEH 83 R. T. YEH: Software Engineering – The Challenges, IEEE Spectrum, Nov. 1983

Structured design methods – how expert systems can help

Peter Eagling
(CCTA UK)

1. Introduction

In recent years the use of structured methods for the analysis and design of commercial data processing systems has grown at such a pace that very few large organisations would now embark on a new computer project without at least considering one or more of these methods.

At the beginning of the present decade, the Central Computer and Telecommunication Agency (CCTA, the UK Government's central computer advisory bureau) commissioned the design of a structured method suitable for use in all Government computer projects. The method was given the title Structured Systems Analysis and Design Method (SSADM), and has now been a mandatory standard in Government for four years.

Currently some 3000 Analysts and Designers in approximately 250 projects are using SSADM.

The CCTA has been very active in the development and installation of SSADM and, more recently, in the provision of computer based tools to support practitioners of the method.

We currently have available a fairly full set of tools enabling all of the documentation, diagrams, forms etc., to be prepared on a high-resolution, micro-computer; our current area of research and development is in the use of expert systems to guide, advise and protect the practitioner as he proceeds through the stages of the method.

The next section of this paper briefly describes SSADM, Section 3 gives an overview of our current tool set and Section 4 shows how we are using an expert system approach both now and in the future to significantly enhance the benefits of computer support for structured methods.

2. Overview of SSADM

2.1 SSADM – The method

The Structured Systems Analysis and Design Method (SSADM) is the standard method for systems analysis and design in Government batch and transaction processing projects. The method takes as its input an initial statement of requirements, and produces as output:

- program specifications;
- user manuals;
- operating schedule;
- file design or database schema;
- plan for testing and quality assurance.

In SSADM, analysis and design are clearly separated activities. Systems analysis is concerned with defining what has to be done, systems design with how it will be done. The complete process is carried out in six phases:

Systems analysis

1. Analysis of current system
2. Specification of required system
3. User selection of sizing and timing options

Systems design

4. Detailed data design
5. Detailed procedure design
6. Physical design control

Each phase is broken down into steps and activities. There are clearly defined interfaces between steps in the form of working documents and criteria for review and project reporting.

Fig. 23 shows how the six stages interrelate.

2.2 SSADM – The techniques

There are three major, diagrammatic techniques involved in the development of an SSADM project, namely:

- Logical Data Structures;
- Dataflow diagrams;
- Entity Life Histories.

Together these form an integrated, three-dimensional view of the data (and associated procedures) of the system, which can be checked, mechanistically, for consistency and completeness. A brief description plus an example of each technique follows.

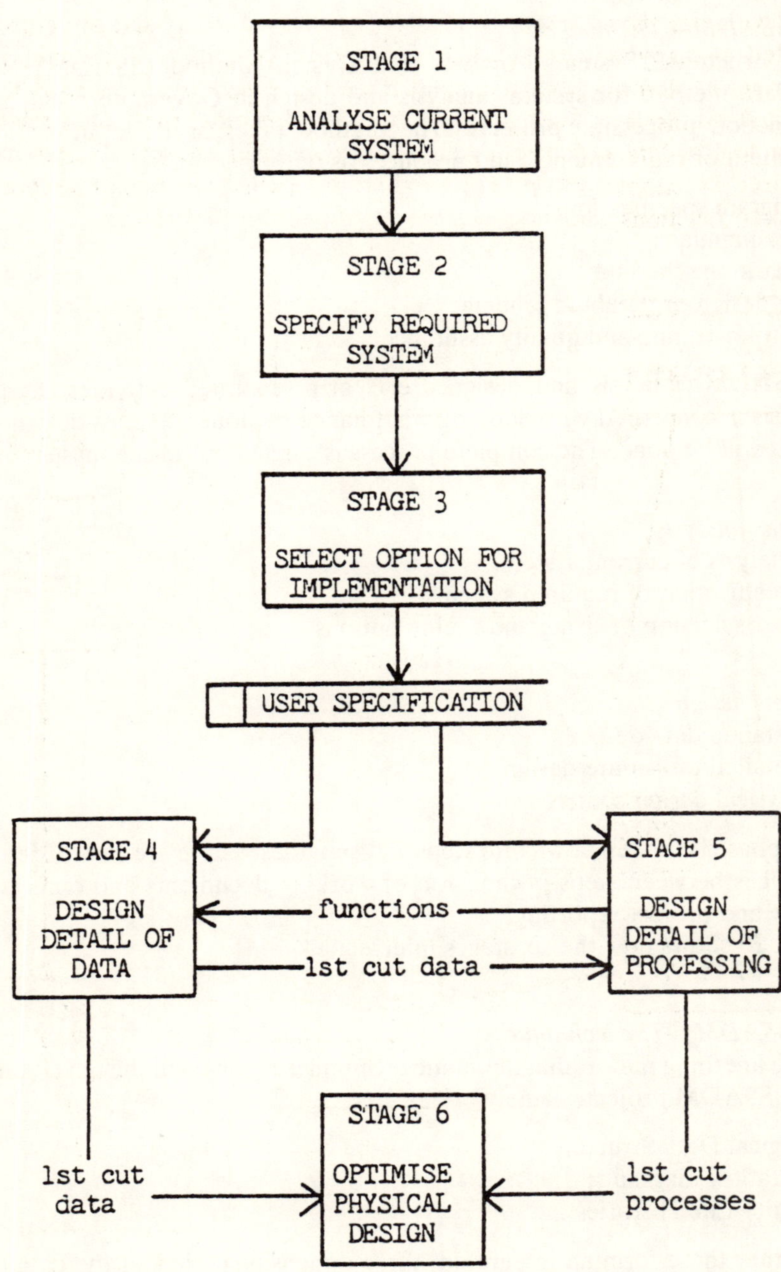

Fig. 23 *Block diagram of SSADM stages*

2.2.1 Logical data structures

The logical data structure is a user view of the business, which is used as a basis for developing the system to support the business. Logical data structures are created at many different points in system development, from initial trial attempts to describe the system in the early stages of investigation, to a detailed synthesis of such a model from third normal form data relations. When logical data structures are built, they are validated by ensuring that the required functions can be supported. If the structures cannot support the required functions, then they are modified in order to do so.

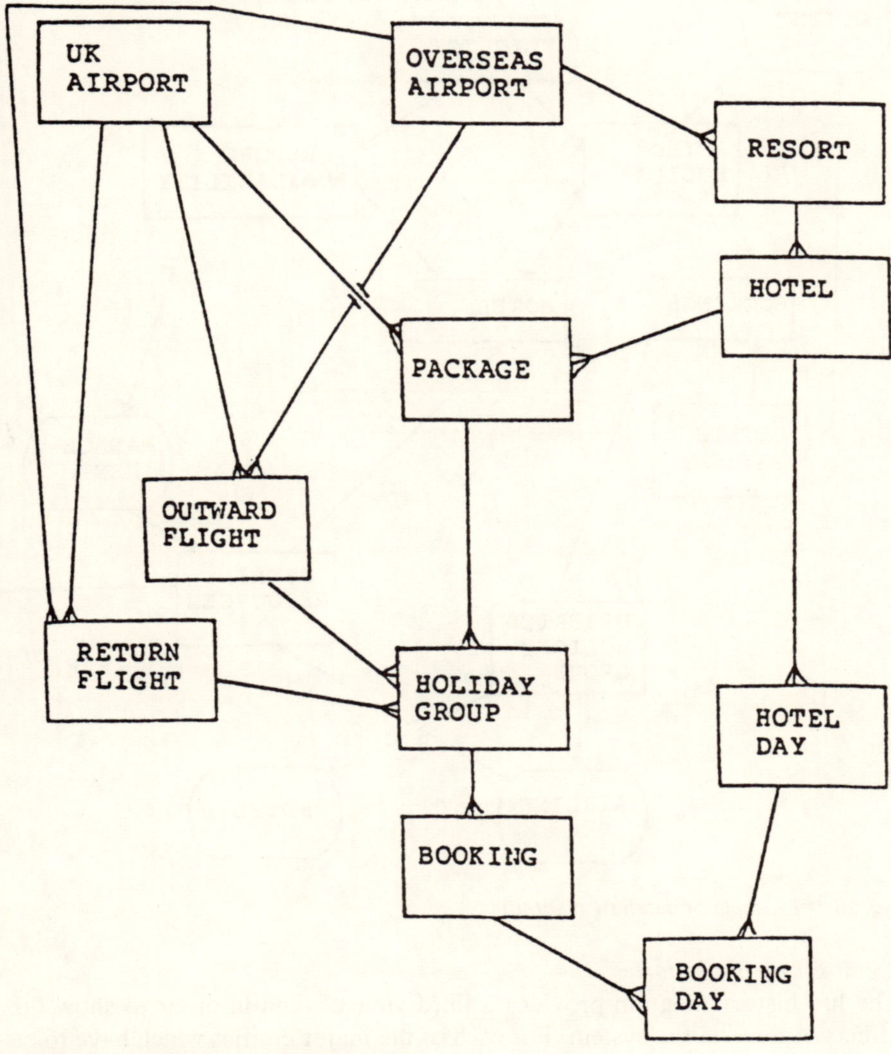

Fig. 24 *Example of logical data structure*

2.2.2 Dataflow diagrams

The dataflow diagram is an operational picture of how data moves around an information system, and moves between the system and the outside world. Whereas the logical data structure is a user view of the business, the dataflow diagram is the user view of the information system. Dataflow diagrams are used to represent existing physical systems, to abstract from these the information flows which support user needs, and to define a logical picture of the required system. After physical design control, the operations schedule is built around the dataflow diagram.

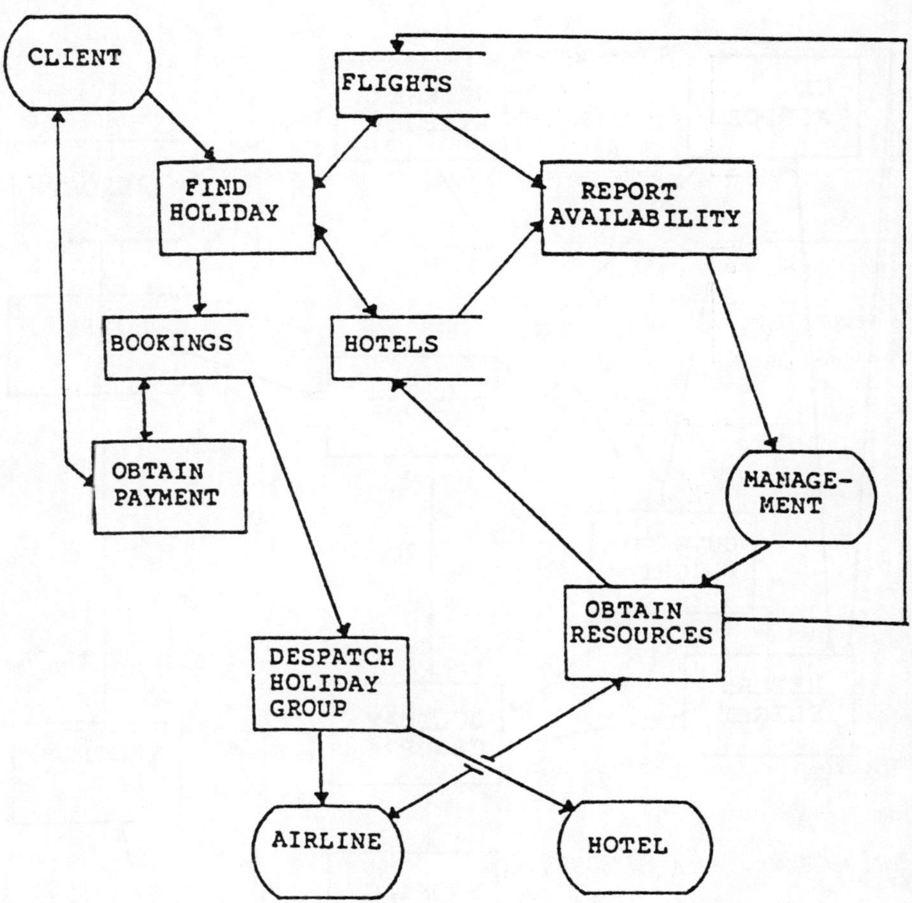

Fig. 25 *Example of dataflow diagram*

2.2.3 Entity life histories

The life history diagram provides a third view of data in order to show the effects of time on the system. For each of the major entities which have to be represented in the system, a chart is drawn showing all the events which affect or update its data in the system. When defining entity life histories the origin of each transaction must be identified, and the processing of each transaction defined. Further detail, such as error recognition and processing and dealing with unexpected events, is then added.

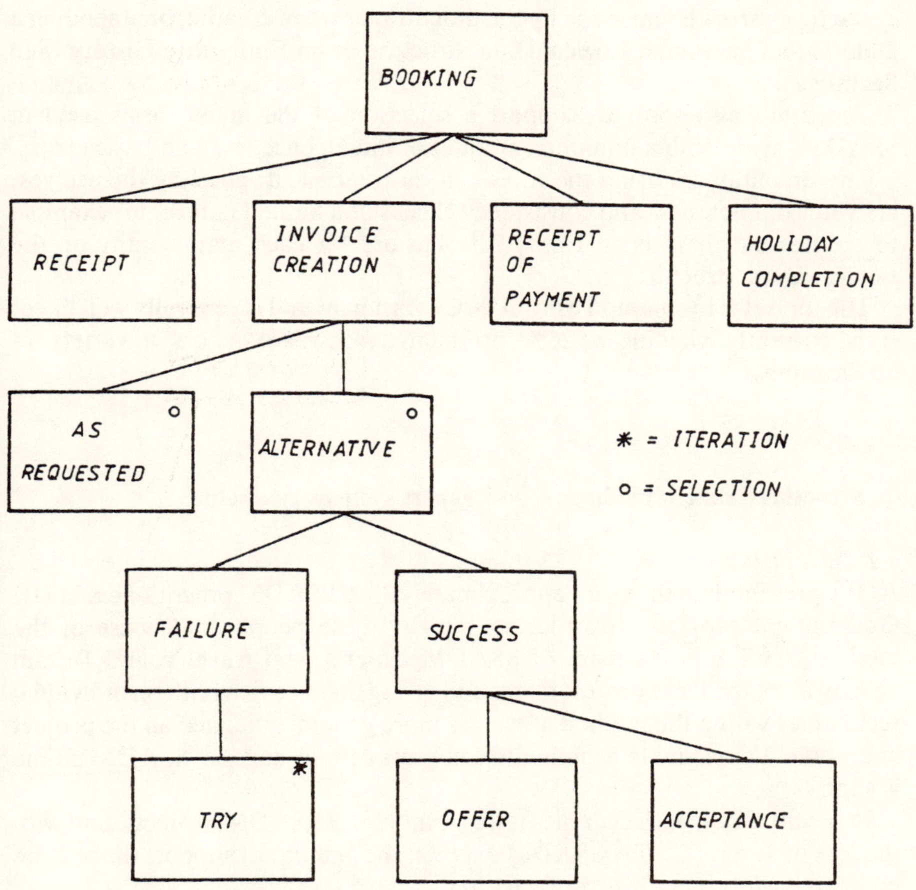

Fig. 26 *Example of entity life history*

3. CCTA's current SSADM toolset

The basic philosophy behind our currently available toolset is to provide sufficient functionality to enable most of the documentation of the method to be prepared on a (micro) computer, together with some checking for completeness and consistency of the diagrams, forms etc. that have been created or amended.

There is very little control over when or how any particular tool is invoked; the practitioner is thus free to use the wrong technique/tool at the point in the method so far attained. There is no advice or guidance given apart from a simple 'help' facility.

The tools run under Multi User Concurrent CPM and are written in a combination of 'C' and Pascal (for historic reasons). There are diagram editors with a fairly simple degree of intelligence in order to avoid the most basic syntactical errors being made by the practitioner when creating or amending a Data Flow Diagram, a Logical Data Structure or an Entity Life History (See Section 2).

There are also tools to support a selection of the many forms used in SSADM, again with a minimum amount of intelligence to avoid basic errors.

The practitioners using the tools can, at intervals decided by themselves, activate completeness and consistency checks of a limited nature, for example to ensure that there is an Entity Life History for each major entity on the Logical Data Structure.

The toolset is in common use in UK Government and is generally well liked. It is currently yielding a 26% productivity increase across a variety of applications.

4. Structured design methods – how expert systems can help

4.1 Introduction

At the present time there are approximately 3000 SSADM practitioners in UK Government projects. In order to support these people in the use of the method, CCTA has a team of SSADM experts who travel round Britain visiting all of the large project teams and giving them advice both on individual techniques within the method as well as more general guidance on the project as a whole. For example, whether to compress certain stages of SSADM if time is short.

In recent times, with ever increasing numbers of SSADM projects, and with the loss of several CCTA SSADM experts, the amount of support that can be given to each project is severely limited.

The question that we have been recently addressing at CCTA is, 'Can expert systems to some extent at least, replace our limited human expert resource?'. In order to answer this question we are proceeding along two separate routes so that we gain experience quickly.

4.2 Database schema design

One approach that we have recently taken is the commissioning of the building of an expert system, to a large software house in the UK, to give advice to SSADM project teams on the transition from a logical specification of the data of the application under construction, to a physical implementation of that specification on an actual DBMS (ICL's IDMS(x)).

Advice on this transition is a common reason that our human experts are called in to a project, and therefore if this experiment is successful, very significant savings in man-power will be achieved.

This system is being written in Prolog and runs on both the Sun workstation and the Torch 68000 Unix system.

The program interacts with a practitioner, asking him questions on the shape of the data specification, the transaction timings required, sizing information etc. and then, using rules previously extracted from human experts, gives advice as to a schema design that is likely to be close to optimal. The program may be interrogated by the user as to why it recommends a particular design or sub-design. However the rules are very complex and the explanations are therefore sometimes very involved. Generally the advice is taken without much argument!

The Prolog program is very large indeed and, as might be guessed, executes fairly slowly (an average advice session takes around two hours including questioning of the practitioner). This is not too bad because the logical to physical transition only has to be made once per project although, of course, several attempts might have to be made to achieve optimum results.

The first version of the program has been demonstrated to us by the software house concerned, and it works well (albeit slowly), further enhancements are now being made to improve the user interface and to increase the execution speed.

4.3 Guidance advice and protection

Section 4.2 gave information on a specific use of an expert system in the latter stages of SSADM. Our second area of investigation is more general and covers the whole of the analysis and design phases of the mehod.

Before we could tackle this line of work, we needed a new vehicle of computer support on which to place the expert system. Very briefly, this vehicle comprises two machine types:

(*a*) one project machine per project on which all project information is held and all project-wide consistency and completeness checking carried out;
(*b*) one personal machine per practitioner on which support tools for the creation and amendment of all the documents of the method are placed.

The unit of work on each personal machine is a document. Documents are the sole means of information exchange between project and personal machines.

The expert system would be placed on the project machine.

The first philosophy of this approach is that each practitioner should be guided, advised and protected (from himself) as he proceeds through the stages, steps and tasks of SSADM. This is the main role of CCTA's human experts during the small amount of time they have available when visiting each project; the advantage of a computer expert is that its vigil will be continuous and complete throughout the project development period.

The second philosophy is that all consistency and completeness checks will be rule based rather than procedural based as with our current toolset.

Indeed, all knowlege and rules of the use and the techniques of SSADM based on the experiences of our human experts is encoded into a knowledge base residing on each project machine.

There are three main repositories of information on the project machine:

(*a*) the knowledge base referred to above;

(*b*) the project database containing all of the logical design information of the project under construction;

(*c*) the document database containing all physical information of each document created and amended during the analysis and design phases of the project.

SSADM is a prescriptive method of stages, steps and tasks, but is nevertheless informal in the mathematical sense. The rules therefore relate to the use of the method and to the individual techniques rather than trying to ensure mathematical rigour in the transition between these stages, steps and tasks.

The expert system is never explicitly invoked but is active all the time that the project machine is switched on. It continually monitors the state of the project database while it is being updated by all the practitioners of a given project, comparing that state with the pre-conditions referred to within the rules in the knowledge base (on the same project machine) to see if any action, guidance or warning should be triggered. Sometimes the action is internal, that is, either the project database or the document database should be updated by the expert system. On other occasions it is necessary to communicate with one or more of the practitioners currently using the system.

This communication can be at one of three levels:

- an error signal, which shows that a technique within the method has been violated;
- a warning, which indicates some degree of incompleteness, albeit only temporary;
- or guidance, which warns the practitioner of deviation from the preferred style of use of the method.

Each of these types is triggered by one or more rules within the knowledge base.

The knowledge base is encoded in a format of our own design, mainly because we suspected that more general techniques/languages would not yield the performance characteristics that we required.

5. Summary

CCTA have invested substantial resource in the development and support of one particular structured design method – SSADM.

Over the past two or three years we have developed a full set of computer based tools of limited intelligence that have been extremely useful to a large number of SSADM users.

With more and more projects to support and with fewer and fewer SSADM experts to give advice, we are investing heavily in the research and development of the incorporation of expert system components into our existing and future SSADM tool sets, in order to maintain, if not improve, the quality and consistency of our advisory service.

Formal methods: present and future

***M. I. Jackson**

Praxis Systems plc, 20 Manvers Street, Bath BA1 1PX, United Kingdom

1. Introduction

It is widely accepted that industrial software production has a number of associated difficulties and problems. These include the following:

The proportion of system costs due to software has increased dramatically over the last twenty years and is continuing to do so. This poses major problems for technical management who are frequently qualified and experienced in more traditional engineering disciplines.

A major skill shortage exists in software and is expected to continue to increase. (For example, it has been estimated by the US DoD that the USA is currently short of 200 000 programmers and will be short of 1 million by 1990.)

The quality of delivered software is frequently inadequate in performance and reliability. For example, a US army study showed that for a 2 million dollar budget, less than 2% of the software was used as delivered.

Software projects are subject to frequent cost and timescale over-runs. The causes are usually inadequate engineering methods to handle the complexity of systems and inadequate project management techniques.

The discipline of software engineering is in its infancy. There is at present relatively little underlying theory or generally accepted good practice on which to base an engineering approach. As a consequence, software systems tend to be 'crafted' rather than engineered.

These (and related) problems have by now received world-wide recognition. A number of major initiatives have consequently been instituted to address these issues, for example, the Alvey Software Engineering programme and the ESPRIT Software Technology programme.

* Previously with: Standard Telecommunication Laboratories Limited, London Road, Harlow, Essex CM17 9NA, United Kingdom.

From an industrial point of view, it is clear that the following global objectives should be addressed:

Increased productivity in software production (to address skill shortages).

Higher quality in software products, e.g. improved reliability, performance and adaptability.

More effective use of human and machine resources, e.g. by automating routine tasks as far as possible thus allowing highly skilled and scarce staff to concentrate on intellectual tasks.

These objectives need to be addressed by developments in a number of areas. The remainder of this paper will concentrate on one area in particular – formal methods.

2. Formal methods

The introduction of formal methods of system development is seen as a major step towards improving the software engineering process. Formal methods are rigorous engineering practices based on mathematically formal foundations. Unlike most of the present day approaches in widespread use, formal methods provide the means of expressing system designs concisely and unambiguously and in such a way that the behavioural aspects of systems can be explored through formal logical manipulation.

The Alvey Software Engineering Directorate has established a major programme in the formal methods area (ALV 84). This programme emphasises three areas of activity: the rapid exploitation of mature formal methods, the industrialisation of promising methods so that they can be exploited in the near future, and fundamental research to provide more powerful methods in the longer term.

The Alvey formal methods programme foresees a significant number of benefits arising from the introduction of formal methods. Most significantly, they are seen as providing the scientific basis underlying software construction. They will allow at least the level of confidence in software designs as in other more established branches of engineering, and will be the key to the certification of software in safety-critical applications. Formal specifications are seen as particularly important for providing a firm contractual basis for the interaction between the supplier of, and the client for, a piece of software. They also provide a natural basis for rigorous interface descriptions thus simplifying the problem of reusing software components. Finally, the use of formal methods early in the life cycle should uncover many specification and design errors which otherwise might not have been detected until system test and operation when their repair would be very (if not prohibitively) expensive.

The Alvey programme characterises a mature method as having the following attributes:

> Books, technical reports and journal articles are widely available and accessible to all sections of the community.
>
> Training programmes have been developed and given field trials in industrial contexts
>
> Industrial case studies have been conducted and the results published. Evaluations of these case studies have also been conducted and published.
>
> The method has had some, perhaps small scale, production experience in industry.
>
> Some useful tool support, possibly of an experimental nature, should exist.
>
> The major strengths and weaknesses of the method are reasonably understood.

These characteristics are rather demanding, but they are the essential minimum that must be achieved before a method can be successfully introduced into an industrial environment. They indicate the significant level of effort that must be devoted to 'industrialising' promising ideas from the research community before they can be assimilated by industry. Currently very little effort is applied to this task, which is one of the major reasons for the gulf between theory and practice. Even if a method is deemed mature according to the criteria given above, much more experience will be required before it achieves widespread adoption in industry.

3. STC experience of VDM

Standard Telecommunication Laboratories, the central research laboratory of STC, has been conducting research into formal approaches to system development since 1979. In 1982, a major new project in the office systems area requested advice on formal specification methods. Of the methods under consideration at the time, one in particular, the Vienna Development Method (VDM) was recommended for consideration.

VDM originated in the Vienna Research Laboratories of IBM and is most closely associated with D. Bjorner and C. B. Jones. The reasons for its recommendation were largely pragmatic, for example:

> Considerable user experience of the method already existed within organisations such as IBM and the Danish Datamatik Center. A number of case studies have been published, for example in the book by Bjorner and Jones [BJO 82]. The strengths and weaknesses of the method are well understood.

Training and consultancy in the method were available. Professor C. B. Jones, now of Manchester University, was willing to present an established two week course to the project team and to provide continual support through consultancy.

A well-written and accessible text book produced by Professor Jones was available to support the course material.

In the summer of 1982 an evaluation exercise was conducted in order to assess the suitability of VDM as a specification vehicle for the proposed project. This exercise was organised as follows:

A subsystem of the project was chosen as the case study to be used.

A small group of analysts, having no previous familiarity with formal methods, were selected and introduced to VDM.

Two STL staff members were asked to support the analysts as consultants (alongside Professor Jones).

An observer, independent of the analysts and consultants, was appointed to identify assessment criteria before commencement of the evaluation, to observe the conduct of the exercise, to write the evaluation report and to produce recommendations to management.

The chief system designer was involved from time to time to answer questions regarding the requirements and to play the role of 'customer'.

The evaluation exercise took as its input an English language statement of requirements. Its output were the corresponding VDM specification, the evaluation report and recommendations for VDM.

In the light of this exercise, STC decided to adopt VDM for the project, and set in hand work in the areas of training, standards and support tools. The method was subsequently adopted for other internal projects.

The experience of VDM can be summarised as follows:

The identification of abstract data structures in the formal specification aids conceptualisation of the eventual product and supports the optimal choice of operations and functions for the eventual users of the system. The iterative evaluation of the specification against user requirements assists dialogue with the customer and continues until a version acceptable to the customer is produced.

The VDM specification language provides a number of useful thinking tools which allow the analysts to distinguish more easily between WHAT the system would do and HOW it should do it. The analyst is also forced to consider many more aspects of the requirements than with previous (non-formal) analysis techniques.

The precision of the specification language reveals many previously unrealised anomalies and inadequacies in the informal statement of requirements. Rigorous reasoning about the specification improves confidence in its adequacy. An improved natural language requirements specification can be produced based upon the formal specification.

As a result of growing interest in VDM within the company, a programme of training courses was developed. These comprise:

A Perspective on VDM. This is a three day course aimed at project managers, team leaders, consultants and support staff from areas such as quality assurance, technical documentation or marketing and who require a sound understanding of VDM. It provides a broad view of formal methods, the ability to read and review VDM specifications, and advise on the introduction of VDM into a project.

VDM Foundation Course. This is a ten-day intensive course for technical staff wishing to produce system specifications written in VDM and for those needing to design systems to satisfy VDM specifications. Students completing the course should be able to read and write VDM specifications, demonstrate that a specification meets a requirement and demonstrate that a design meets a specification.

These courses are now offered to the public by STC which operates a commercial software and training service.

Various language development activities for VDM have also been conducted in STC. In the short term, it was recognised that there was an immediate need for facilities to allow the specification process to be carried out 'in the large' and for designs to be documented alongside the specifications that they implement. STC consequently undertook the design of an extended design language which included a module construct and a design pseudocode besides the essential components of the VDM specification language. (This work is heavily influenced by work undertaken at IBM's Boblingen laboratory on the SLAN-4 language [BE 183].) This language is supported by a UNIX-based syntax checker.

In the longer term, it was recognised that in order to extend the VDM specification language in a coherent way, and in order to build more powerful automated support tools, it would be necessary to improve the standard of definition of the existing specification language. Consequently, a joint activity was started by STL and Manchester University to define for STC a VDM Reference Language to be used as the STC standard. To date, documents describing the concrete syntax, abstract syntax, type model and context conditions of the Reference Language have been issued and further work on semantics is underway. In addition, it is intended to develop a more extensive support toolset around the Reference Language, and a UK Alvey project has recently commenced for this purpose. Recently, work has begun to develop a

UK Standard Definition of the Reference Language based on the work carried out in STC.

In conclusion then, the STC experience indicates that, as a method for formal specification, VDM can be successfully introduced, if appropriate levels of investment are made in the areas of:

Evaluation, by case studies, of the suitability of the method for the application area of interest. The valuation criteria should be properly defined, monitored and assessed.

Training for staff, of a professional quality.

Consultancy, as appropriate to particular needs, by skilled and experienced personnel.

General support, for example, by developing and implementing corporate standards.

In the longer term, support tools, such as a specification-oriented database and a syntax-directed structure editor.

For more information on VDM, readers are referred to the book by C. B. Jones (JON 80]. Readers requiring further details of STC's experiences with VDM are referred to a recent paper [JAC 85].

4. Formal methods in the life cycle

There are many ways of viewing the software life cycle, the best-known being the standard 'Waterfall' method. In the alternative 'Contractual' view of the life cycle, each phase is regarded as involving two parties, one fulfilling a 'customer' role, the other fulfilling a 'supplier' role. So, for example, at the requirements specification phase, the customer role will be filled by the real customer for the system, while the supplier will be the analyst whose function is to supply the formal specification. The customer will provide an initial informally worded requirements statement and the analyst will produce a first attempt at a formal specification which he first verifies for internal consistency and completeness. (In effect he asks 'Is this a specification of some system even if it is not quite what the customer wants?'.)

Having produced this first attempt, he must then demonstrate the effects of the specification to the customer. It is obviously unreasonable to expect the average client to be able to read the formal anotation directly, though there are classes of (more sophisticated) customer where this does not apply. In general the analyst will use a process known as 'animation' to demonstrate the effects and consequences of the specification. This might involve a symbolic execution technique, for example, by moving tokens around a Petri net, or might involve the interpretation of behaviour derivable by theorems from the formal specification in the client's domain of knowledge. Inevitably, the first specification will not match the customer's precise requirement so further iterations of this process will be necessary. Eventually, a specification will be produced to which the customer agrees and this can be forwarded to the design stage (of course, problems may be discovered later which cause the requirements to be reconsidered.

In the design phase, the analyst becomes the 'customer' and the designer 'supplier' with the responsibility for producing a design which is consistent with the specification. Once again, the processes of verification and validation are iterated until an acceptable design is produced.

These contractual relationships can continue through however many design and implementation phases as are necessary.

The contractual model presents a somewhat idealised view of software development, but is useful as a basis for examining how formal methods might be used. There is general agreement that formal methods should be used initially at the specification phase since the greatest cost savings will be made by locating specification errors early. At the design and implementation phases, the degree of formality used would probably be less, since, except in the case of very critical projects, the costs would probably outweigh the benefits. In most cases, a rigorous approach to design would be more acceptable, with the verification and validation techniques being more informal and traditional in flavour. Testing strategy could, perhaps, be based upon the formal specification. Post-developmental phases, such as maintenance or enhancement, would be considerably assisted if formal specification were kept in the project database since the implications of proposed changes could be more easily assessed.

Training and education are areas which require attention. It must be emphasised that the introduction of more formal methodologies does not reduce the skills needed to develop software. On the contrary, the newer methods require a different set of skills to traditional programming activities. Higher standards of education and professionalism are required. Courses must be developed (some already exist for the established methods) to provide analysts and designers with the necessary skills in particular methods. Managers, too, need training to manage reviews and monitor progress of projects using formal methodologies. Educational courses must be provided to give the necessary background in discrete mathematics which many experienced practitioners lack (through the relevant material is gradually being included in many undergraduate computer science courses).

5. Automated support for formal methods

The most basic type of support tool is one that supports syntax and type checking. For any well-defined specification language a syntax and type checker can often be provided using standard parser generators. Coupled with a reasonable file or database system, such a tool would provide an invaluable aid to the system specifier and should be relatively cheap to provide. Other syntax-oriented tools include syntax-directed editors, pretty printers etc.

A more sophisticated type of tool would support symbolic execution of specifications and designs. The difficulty of building such tools depends to some extent on the formal method being supported, but clearly such an aid would be most useful as a validation and animation tool.

Validation would also be supported by theorem proving aids which support formal verification of specifications and transformations during the various phases of design. Recently, there has been a trend away from large theorem provers which consume large amounts of computing resources to interactive systems which assist the human in carrying out a proof. Related is the work on automated program transformation which promises to assist the designer in producing correct implementations with regards to the specification.

The underlying support environment cannot be ignored. Developments such as the Ada APSE or the European PCTE promise much as the central database will be provided. If a proper database were provided, it should be possible to build libraries of reusable software components each having a formal specification. By composing specifications of existing parts, and by integrating them with the specifications of new parts, it should be possible to demonstrate that a larger system specification could be met. Thus, system development might become a more bottom-up activity with greater reuse of previous work than is possible at present.

Prototype versions of most of the tools mentioned above are already available, at least in research environments. It is reasonable to expect that production quality versions will appear over the next few years.

6. Conclusions

The next few years will witness an increasing use of formal development methodologies as users become more sophisticated and attempt to detect errors in the specification and design phases of the life cycle. Formal methods are already worthy of serious consideration as a specification aid and a rigorous design framework has also been shown to be highly practicable. The results of major research projects established under programmes such as Alvey and ESPRIT, will lead to more extensive and adaptable methods for future use, with a greater level of automated support.

Readers wishing to obtain a more detailed view of the current state of various approaches to formal methods are referred to the comprehensive survey by Cohen et al (COH 86].

7. References

[ALV 84] Alvey Programme Software Engineering. Programme for formal Methods in System Development, Alvey Directorate, April 1984

[BEI 83] BEICHTER, F., HERZOG, O., PETZSCH, H. SLAN-4: A language for the Specification and Design of Large Software Systems, IBM Journal of Research and Development, Vol. 27, No. 6, November 1983

[BJO 82] BJORNER, D., JONES, C. B. Formal Specification and Software Development, Prentice Hall, 1982

[COH 86] COHEN, B., HARWOOD, W. T., JACKSON, M. I. The Specification of Complex Systems, Addison Wesley, 1986

[JAC 85] JACKSON, M. I., DENVIR, B. T., SHAW, R. C. Experience of Introducing the Vienna Development Method into an Industrial Organisation. Procs. Int. Conf. on Theory and Practice of Software Development [TAPSOFT), Berlin, March 1985. Published as Springer Verlag LNCS 196

[JON 80] JONES, C. B. Software Development – a Rigorous Approach. Prentice Hall, 1980

Animation of requirement specifications

Giuliano Pacini, Franco Turini
Dipartimento di Informatica, Univ. di Pisa

1. Introduction

Rapid prototyping has been recently advocated as a possible solution to the problems of poor flexibility of the traditional software life cycle [ACM 82, ALA 84, BAL 83, BUD 84, DEA 83, SMI 85, WAS 82, ZAV 84].

Indeed, the possibility of animating the system since the requirement analysis phase can point out, early in the software construction process, misunderstandings between final users and software producers.

The sentence *rapid prototyping* has been given several meanings. In this paper the meaning is identified with the concept of *executing requirement specifications*.

The execution of requirement specifications presents many crucial points, some of which are:

- The specification language must be, at the same time, non procedural, in order to be amenable for the specification of requirements and executable, to be amenable for the animation of requirements.
- The concept of execution has to be generalised. Indeed one does not want to execute requirements in the usual sense, but rather to check whether the system does or does not have certain desirable properties.

The paper describes the approach developed in the ESPRIT funded ASPIS project [ASP 85]. The main ingredients of the ASPIS recipe are:

- Using a logic based specification language capable of expressing both data flow aspects and functional aspects.
- Translation of the specifications into an extended logic programming language [KOW 79]. The extension is needed for coping with partial functional specifications.
- Exploitation of a generalised logic programming interpreter (Prolog interpreter [CLA 84, CLO 81]) for animation purposes. The flexibility of Prolog interpreters (executions as proofs, logic variables etc.) offers the basis for the required generalization of the execution process.

The paper is organised as follows: Section 2 outlines the specification language. Section 3 discusses a generalisation of logic programming and Section 4 presents the translation of specifications into the extended logic programming and the way the resulting logic program is exploited for animation purposes.

2. RSLogic: a requirement specification of formalism

Usually, the distinction between the requirement specification phase and the design phase [SOM 82] is explained saying that the former has to do with *what* the system has to perform, whereas the latter has to do with *how* the system has to perform the *what* specified during the requirement analysis phase.

In ASPIS, the requirement analysis is based on the SADT methodology [RSC 77, ROS 77]. The SADT methodology is essentially based on a special class of data flow diagrams (SA diagrams), which allow to specify both the functions performed by the system in terms of labelled boxes with input and output ports, and data flow among them in terms of connecting arcs among boxes.

Indeed there is much more than that in the SADT methodology, e.g. an entity relationship based technique for specifying data, but for the purpose of this paper the functional and data flow aspects are sufficient.

In order to integrate the informal aspects of SA diagrams with formal specifications, a novel formalism, termed RSLogic, has been designed. The principal specificity of the formalism, which is logic based, is its ability of taking into account the data flow information implicit in a SA diagram. In SA diagrams semantic aspects are reduced to names, which can label both boxes and connections. Box labels give the name of the functions associated to the boxes; connection names denote the types of the data flowing through the connections.

The envisaged formalism must allow to reason in terms of boxes, box inputs, box outputs, rules to determine output values from input values and so on. To this end, RSLogic allows to write axioms in terms of functions, predicates and *positions* in the diagram. Moreover, the validity of properties of the data, either in input or output to boxes, are in general depending on the paths previously traversed by the data in the diagram (*histories*). So, in RSLogic axioms, positions can be collected into histories. This kind of view point is typical of modal logic. In model logic the truth of logic assertions is not an absolute matter. In modal logic a set of *situations* (ways of existence of the world) is assumed. Truth of assertions is relative to situations. In pure modal logic, nothing is said or imposed on situations. However, modal logic can be specialised, by interpreting situations. For instance, in temporal logic, situations are instants of time. To reason about SA diagrams, the principles of modal logic can be exploited, in that *histories* can be considered situations.

RSLogic must be able to manage the refinement notion which is basic in SA method. A given SA specification (level n + 1) is the refinement of another specification if each box of level n has been expanded in a whole diagram at level n + 1. To establish the refinement some correspondence between the objects of the two levels must be precise. In agreement to the RSLogic approach such correspondence can be stated as follows.

(1) A set of positions at level n + 1 is associated to any position of level n.
(2) An expression composed by functions of level n + 1 is associated to functions of level n.
(3) A boolean expression composed by predicates of level n + 1 is associated to predicates of level n.

The idea is that, when appropriate refinement declarations have been given, as outlined above, the semantic axioms of level n can be automatically translated to level n + 1, i.e. they are rephrased in terms of positions, functions and predicates of the refining level. The refinement is correct if the axioms of level n + 1 implies the axioms of level n, as they have been translated to level n + 1.

When boxes have reached a suitable level of simplicity, functions will be specified by axioms not containing positions. In other words by first order (or Prolog) axioms.

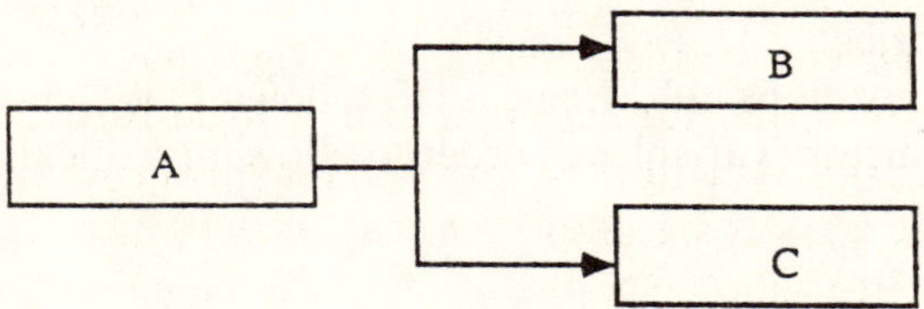

Fig. 27

2.1. Positions in SA diagrams

Intuitively, histories record movements in the diagram, for instance in terms of traversed boxes. However, in many cases, references to positions in the diagram (typically, entry points or exits of boxes) are convenient. For example, consider the structure of Fig. 27 and suppose box A outputs some data a, part of which is routed to B and part to C.

The fact may be described saying that if you reach the exit of A then you will reach the entries of B and C, possibly stating how b and c are extracted from a.

So, the RSLogic proposes the concept of position and allows to describe SA diagrams as organised sets of positions. RSLogic axioms containing references to positions will allow a joint description of structural and semantic aspects.

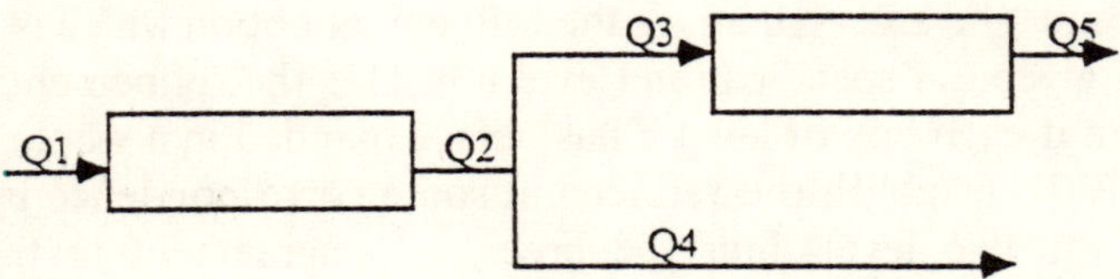

Fig. 28

2.2. Histories

A history is defined as a sequence of sets of (position, value) pairs.

The characterisation of histories as sequences of sets allow handling parallelism in histories. Consider for instance the structure of Fig. 28. A possible history is:

$$\{<Q_1,V_1>\}\|\{<Q_2,V_2>\}\|\{<Q_3,V_3>,<Q_4,V_4>\}\|\{<Q_4,V_4>,<Q_5,V_5\}$$

which can be interpreted as Q_2 gives rise to Q_3 and Q_4, then Q_3 produces Q_5 while Q_4 remains active.

Intuitively, while traversing a box, some input data are consumed while other data are produced in output. From this view-point, it is quite natural to think that data are located in terminal positions of histories only. We will say that a subset of positions are *active* at any time, i.e. some data are located in these positions. The subset of active positions coincides with the last set of positions in histories.

The syntax of positions in RSLogic is the following. We introduce position, position set and history variables in order to allow quantification on positions and histories.

Alphabet

$q_1,q_2,q_3, \ldots ,q_i, \ldots$	position constants (q)
$r_1,r_2,r_3, \ldots ,r_i, \ldots$	position variables (r)
$s_1,s_2,s_3, \ldots ,s_i, \ldots$	(position, value) set variables (s)
$h_1,h_2,h_3, \ldots ,h_i, \ldots$	history variables (h)

Syntax

Postions (pos)

q, r

Position-Value Pairs (pv)

<pos,exp>

Position-value pair sets (pvs)

s, $\{pv^1,pv^2, \ldots ,pv^j, \ldots\}$, $pvs^1 \cup pvs^2$, $pvs^1 \cap pvs^2$

Histories (hs)

h, pvs, $hs^1\|hs^2$

Position atomic fomulas (pafm)

$pos^1 = pos^2$, $hs^1 = hs^2$, posε pvs

RSLogic considers expressions and atomic formulas in which positions sets are mentioned (pafm). It is possible to compare, join and intersect position sets. These kind of phrases, as well as the quantification on positions, position sets and histories, aim to provide large flexibility in the role of histories in RSLogic.

Logic history symbols are pre-interpreted in RSLogic. This pre-interpretation precises the kind of specialisation we are imposing on pure modal logic.

2.3. Viability of Histories

Essentially, the definition of histories is a syntactic definition. Histories are defined by a context-free grammar. Now, it is intuitive that histories can exist which cannot be actually followed in agreement with a given SA diagram. Consider again Fig. 28, the history:

{<q1,v1>}||{<q2,v2>}||{<q3,v3>}||{<q4,v4>}

cannot be followed for structural reasons.

There is a second reason causing not *viability*. A history can be not viable for semantic reasons. Suppose that the semantics of boxes are defined in pre-post-condition style. (As we see in the sequel, pre-post-conditions can be stated quite naturally in RSLogic). Suppose that a precondition (typically a property on the data located in q^3) is associated to q^3. The viability of history

{<q1,v1>}||{<q2,v2>}||{<q3,v3>}||{<q4,v4>}

depends on the data v3 located in q3, that can depend in turn on the initial input. For some inputs the history may be viable while for other inputs it may be not.

Viability is a semantic matter, not a mere syntactic matter. Thus, in RSLogic a function *Viab* is introduced:

Viab: histories → {True, False}.
Viab gives True as a value if the history is viable.

The syntactic aspect of viability is very simple:

Viab(hs)

is an atomic formula of RSLogic. This allows definition of viability of histories by axioms in RSLogic. Essentially, viability is one of the possible semantic aspects to consider in the description of a system.

From a technical viewpoint, Viab is a function whose type is pre-established. In other words, in any possible interpretation of the logic symbols, Viab must be regarded as a function from histories to Boolean.

2.4. Data Description

In general, the description of a system will contain the definition of data manipulated in the system itself. The description of data is an exception in our modal logic, since data and their properties do not depend on histories.

This fact may be reflected in the syntax by the introduction of particular logic symbols (*invariant* symbols). The syntax stresses that the interpretation of such symbols does not depend on histories. In essence, data are described by classical first order logic axioms, involving invariant symbols only (invariant constants, functions and predicates). For instance, the style of definition could be an axiomatic abstract data type style.

The syntax of data definition is the following:

Alphabet

$ic_1, ic_2, ic_3, \ldots, ic_i, \ldots$	invariant constants (ic)
$if_1, if_2, if_3, \ldots, if_i, \ldots$	invariant functions (if)
$ip_1, ip_2, ip_3, \ldots, ip_i, \ldots$	invariant predicates (ip)
$x_1, x_2, x_3, \ldots, x_i, \ldots$	variables (x)

Syntax

Invariant Expressions (iexp)

x, ic, if($iexp^1, \ldots, iexp^j, \ldots$)

Invariant Atomic formula (iafm)

ip($iexp^1, \ldots, iexp^j, \ldots$)

2.5. History Dependence

RSLogic allows to associate values to positions. They allow modelling the flowing of data. Indeed, the state of the system is characterised, first of all, by data flowing in the system. However, in many cases, the state has another component, which has the nature of a global component. This component is made up of information, which are not located in positions, but whose arrangement can be sensible to box traversing. For instance: in a business system, the state depends on data flowing in the system as well as on a (centralised) data base.

To this aim, RSLogic provides *state* symbols (constant, function and predicate state symbols) that contribute to form history depending expressions and formulas.

The syntax of history dependent expressions and atomic formulas is the following. Positions and state logic symbols show their dependence on histories.

Alphabet

$sc_1, sc_2, sc_3, \ldots, sc_i, \ldots$	state constants (sc)
$sf_1, sf_2, sf_3, \ldots, sf_i, \ldots$	state functions (sf)
$sp_1, sp_2, sp_3, \ldots, sp_i, \ldots$	state predicates (sp)

Syntax
Expressions (exp)
sc: hs, sf(exp^1, . . . ,exp^j, . . .): hs, if(exp^1, . . . ,exp^j, . . .)

Atomic formula (afm)
sp(exp^1, . . . ,exp^j, . . .): hs, ip(exp^1, . . . ,exp^j, . . .)

The notation
logic-item: hs
indicates the history dependence.

For instance: 'sf(a): hs' means the value of the state function sf after the history hs. Note that invariant items can take part in history depending formulas. For instance, 'if(. . . ,exp^j, . . .)' can depend on histories because some exp^j can do so.

2.6. Specifications and queries
Structure of SA diagrams can be easily described in RSLogic. In RSLogic we may give axioms in which both structural and semantic aspects are considered. The axiom:

$$\begin{aligned}\forall\ x,s\ & \mathrm{Viab}(x\|\{<q1,v1>\} \cup s) \wedge \\ & \mathrm{fm1}: x\|\{<q1,v1>\} \cup s \\ & \rightarrow \\ & \mathrm{Viab}(x\|\{<q1,v1>\} \cup s\|\{<q2,v2>\} \cup s) \wedge \\ & \mathrm{fm2}: x\|\{<q1,v1>\} \cup s\|\{<q2,v2>\} \cup s\end{aligned}$$

states a transition rule, in terms of pre-post-conditions, from position q1 to position q2, which may be (for example) the entry point and the exit of a box respectively.

RSLogic seems to be able to express the major part of the queries, that can be conceived in order to analyse the properties and the behaviour of SA diagrams augmented with semantic specifications.
Examples:

(a) $\forall\ x\ \mathrm{fm}: x\|\{<q,v>\}$
(b) $\exists\ x\ \mathrm{fx}: x\|\{<q,v>\}$
(c) $\neg\exists\ x,s\ \mathrm{Viable}(x\|\{<q1,v1>,<q2,v2>\} \cup s)$
(d) $\exists\ x,y\ \mathrm{Viab}((x\|\{<q1,v1>\}\|y\ \|\{<q2,v2>\}$.

(*a*) requires the validity of an invariant assertion: fm must be always True in q, i.e. after any history ending in q.
(*b*) is a fairness requirement: the existence of an history (ending in q) with fm True is required.
(*c*) is a mutual exclusion requirement: it must not exist in an history in which q1 and q2 are active at the same time.

Various other kinds of queries are possible. For instance, (*d*) requires that a history exists that touches q1 and q2.

3. Logic Programming with constraints

Logic programming *per se* presents a number of suitable characteristics as an executable specification language:

- It is declarative, i.e. a function is described stating its properties rather than an algorithm for computing it.
- The execution of a logic program is a proof of a theorem, which is derived from the program itself considered as a set of axioms. From a practical viewpoint, this amounts to the possibility of executing the program in many different ways and, ultimately, to the possibility of checking different properties of the program.

In order to achieve an even greater generality of execution of logic programming, the language has been extended with a construct which allows to state constraints both on the variables of goals and the variables of clauses. Such constraints are handled by a generalisation of the executor.

The resulting execution is a form of symbolic execution, i.e. an execution able to handle both concrete values, unconstrained variables and constrained variables.

To extend logic programming in the direction of symbolic execution, we introduce a **where** construct of the kind

where list-of-atoms

The syntax of logic programming is modified to allow **where** constructs in the right part of Horn clauses and in goals.
Examples:

P(f(X,Y)):— Q(X,Y) **where** R(X),S(Y)
:— P(X,Y) **where** K(X)

The elementary inference step of the interpreter is modified in order both to exploit constraints for finishing the proof and to specialise them.

3.1. Exploitation of constraints

Constraints are assumed as *true* properties of the involved variables. In this respect, given a constrained resolvant, the first step is to see whether some of the subgoals are established by the constraints themselves.

Formally let $A[X_1, \ldots, X_n]$ denotes a generic atom predicating on a subset of variables $X_1, \ldots, X_n$.
Let

$$G_1[X_1, \ldots, X_n, Y_1, \ldots, Y_m], \ldots, G_k[X_1, \ldots, X_n, Y_1, \ldots, Y_m]$$
$$\textbf{where } C_1[X_1, \ldots, X_n], \ldots, C_1[X_1, \ldots, X_n]$$

be a constrained resolvant, where the X_i's are the constrained variables.

Then, each goal G_i is unified against each constraint C_j. If G_i and C_j unify with m.g.u. θ then the resolvant is rewritten as follows

$$(G_1[X_1, \ldots ,X_n,Y_1, \ldots ,Y_m], \ldots ,G_{i\text{-}l}[X_1, \ldots ,X_n,Y_1, \ldots ,Y_m],$$
$$G_{i+11}[X_1, \ldots ,X_n,Y_1, \ldots ,Y_m], \ldots , G_{k1}[X_1, \ldots ,X_n,Y_1, \ldots ,Y_m])\ \theta$$
$$\textbf{where}\ (C_1[X_1, \ldots ,X_n], \ldots ,C_1[X_1, \ldots ,X_n])\ \theta$$

The unification step between a goal and a constraint obeys the restriction that a variable occurring in a constraint can only unify with either itself or with an unconstrained variable. This is essentially due to the fact that constraints cannot be considered general statements but their meaning is to refer to specific variables in the context of either the goal or the clause they occur in.

3.2. Specialisation of constraints

Let

$$H[X_1, \ldots ,X_n,Y_1, \ldots ,Y_m]{:}\!-\ S_1[X_1, \ldots ,X_n,Y_1, \ldots ,Y_m], \ldots ,$$
$$S_h[X_1, \ldots ,X_n,Y_1, \ldots ,Y_m]$$
$$\textbf{where}\ K_1[X_1, \ldots ,X_n], \ldots ,K_k[X_1, \ldots ,X_n]$$

be a constrained clause.

Let

$$G[Z_1, \ldots ,Z_p,W_1, \ldots ,W_r]$$
$$\textbf{where}\ C_1[Z_1, \ldots ,Z_p], \ldots ,C_1[Z_1, \ldots ,Zp]$$

be a goal. Suppose that the goal unifies with the head of the clause with m.g.u. θ. Then the goal is rewritten as

$$(S_1[X_1, \ldots ,X_n,Y_1, \ldots ,Y_m], \ldots ,S_h[X_1, \ldots ,X_n,Y_1, \ldots ,Y_m])\ \theta$$
$$\textbf{where}\ (C_1[Z_1, \ldots ,Z_p], \ldots ,C_1[Z_1, \ldots ,Zp],$$
$$K_1[X_1, \ldots ,X_n], \ldots ,K_k[X_1, \ldots ,X_n])\ \theta$$

In words, both the constraints of the goal and the constraints of the clause have been specialised using the m.g.u. The constraints of the new resolvant are the conjunction of the two sets of contraints.

3.3. Answers to queries

According to the outlined extension the system replies to a query with answers of the form

VAR = TERM **where** list-of-atoms

That is the terms in the answers are constrained by a set of atoms, which can contain variables of the original query and, possibly, new ones.

3.4 Constraint evaluation

In order to get significant and general answers, evaluation of constraints is convenient. Evaluation of constraints is possible if the predicates exploited in stating constraints are defined by clauses in the program. Consider, for instance, the following case. Assume that the program contains the clause:

P(f(X)):— Q(X)

and that the current resolvant is constrained by

P(f(Z))

Then, if the above clause is the only one unifying with P(f(Z)), then the constraint can be replaced by Q(Z). Such a rewriting simplifies the constraint and allows a greater generality in the answer. Indeed if a subgoal like Q(Z) will ever be generated, Z will not be rewritten. On the other hand, if the rewriting is not performed, a subgoal like Q(Z) requires the enumeration of all the elements satisfying Q. Currently the system exploits the outlined strategy, i.e. a constraint is rewritten if there is only a clause unifying with it and if the rewriting provides a more general constraint. It is a matter of further study looking for a better constraint evaluation strategy.

3.5. Symbolic execution as a tool for proving implications

Any constrained answer of the system can be considered an implication in that the meaning of the answer can be viewed as

constraints **imply** original query

For example, constraints can be used for expressing type constraint in the axiomatisation of an abstract data type. For example, the answer to the query

eq(car(cons(Z,V)),Z)

is

Z = Z
V = V
  where atom(Z),list(Y)

which can be viewed as

atom(Z), list(Y) ⇒ eq(car(cons(Z,V)),Z)

4. Mapping of requirement specifications into logic programming with constraints

This section describes the animation of requirements specified in RSLogic. Specifications in RSLogic allows to express transitions of boxes and data flow among boxes.

Fig. 29

For example the specification of the box depicted in Fig. 29 could be:

$$\text{viab}(H \| \{<Q_1,V_1>,<Q_2, V_2>\} \text{ and pre}(V_1,V_2) \Rightarrow$$
$$\text{viab}(H \| \{<Q_1,V_1>,<Q_2>\{\| \{<Q_3,V_3>\}) \text{ and post}(V_1,V_2,V_3)$$

where H is a history variable, Q_1,Q_2,Q_3 identify positions in the SA diagram and V_1,V_2,V_3 are the terms denoting the values associated to the positions. It is evident that the axiom is a transition axiom in terms of pre and post conditions.

Since, the aim is to provide an environment where descriptions can be animated, it is necessary to be able to translate RSLogic axioms into a form closer to logic programming.

To this purpose, a translator has been realised, which translates RSLogic axioms into constrained logic programming axioms. The resulting Horn clauses are such that sets of (position, value) pairs can be handled by a special unification process, which has been defined in Prolog itself.

Furthermore, only SA diagrams at the lowest level of refinement are such that the boxes denote reasonable simple functions. In other words, only after some refinements functionalities of boxes are actually axiomatised by Horn axioms. Indeed, at the intermediate levels, some predicates or functions used in the specifications could be not yet axiomatised. For this reason the standard Prolog interpreter is not able to animate the specifications of an intermediate level.

Fig. 30

Let us see, by means of an example, how incomplete specifications can be handled exploiting the **where** feature discussed in Section 3. Consider the SA diagram of Fig. 30 along with the axiom

$$\text{viab}(H \| \{<I,V_I>\}) \Rightarrow \text{viab}(H \| \{<I,V_I>\} \| \{<O,V_O>\}), \text{fun}(V_I,V_O)$$

and suppose that fun is not axiomatised. In this case the translation yields the following Prolog clause

$$\text{viab}(X \| H)\text{:—} \text{ unify}(\{\},\{<I,V_I>\},X),$$
$$\text{viab}(H \| \{<I,V_I>\} \| \{<O,V_O>\}) \textbf{ where } \text{fun}(V_I,V_O)$$

The extended executor, if queried with

$$\text{viab}(H \| \{<O,V_O>\})$$

would reply

YES
$H = H1 \| \{<I,V_I>\}$ **where** $\text{fun}(V_I,V_O)$

As the example shows, the extended execution mechanism allows to query also partially axiomatised specifications, providing, nevertheless, information about the flow of data.

5. Examples

In general, given a box like the one of Fig. 30, it is axiomatised by a transition axiom of the form

$$\mathrm{viab}(H \| \{<I,V_I>\} \text{ union } S), \mathrm{Pre}(V_I)$$
$$\Rightarrow \mathrm{viab}(H \| \{<I,V_I>\} \text{ union } S \| \{<O,V_O>\} \text{ union } S), \mathrm{Post}(V_I,V_O)$$

where S denotes the rest of active positions. Notice the instrinsic asynchrony of the axioms given by the fact that S remains the same before and after the traversal of the box.
The transition axiom is translated into two Horn axioms:

(1) $\mathrm{viab}(H \| \{<I,V_I>\} \text{ union } S \| \{<O,V_O>\} \text{ union } S)$:—

$\mathrm{viab}(H \| \{<I,V_I>\} \text{ union } S), \mathrm{Post}(V_I,V_O)$

i.e. position O can become active after position I if
- (a) position I is *correctly* reached
- (b) V_O is a correct value for position O

(2) $\mathrm{viab}(H \| \{<I,V_I>\} \text{ union } S)$:— $\mathrm{viab}(H), \mathrm{Pre}(V_I)$

i.e. position I can become *correctly* active if
- (a) there is an history leading to I
- (b) V_I is a *correct* for position I

What follows is the application of the animation strategy to a couple of toy examples with the aim of providing the reader with the feeling of the kind of questions can be asked to the system and the kind of answers which can be got.

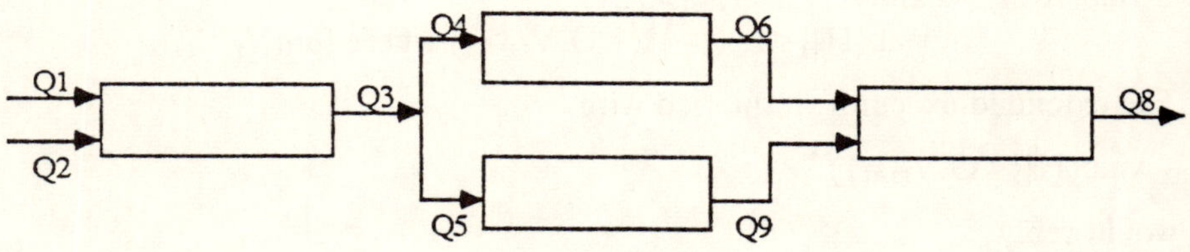

Fig. 31

Consider Fig. 31. The Horn axioms which specify the data flow in it are:

viab({<Q_1,V_1>, <Q_2,V_2>}):—

viab(H || S union {<Q_1,V_1, <Q_2,V_2>} || S union {<Q_3,V_3>}:—
viab(H || S union {<Q_1,V_1>, <Q_2,V_2>}).

viab(H || S union {<Q_3,V_3>} S union {<Q_4,V_4>}:—
viab(H || S union {<Q_3,V_3>).

viab(H || S union {<Q_3,V_3>} || S union {<Q_5,V_5>}:—
viab(H || S union {<Q_3,V_3>).

viab(H || S union {<Q_4,V_4>} || S union {Q_6,V_6>}:—
viab(H || S union {<Q_4,V_4>).

viab(H || S union {<Q_5,V_5>} || S union {<Q_9,V_9>}:—
viab(H || S union {<Q_5,V_5>).

viab(H || S union {<Q_6,V_6>} || S union {<Q_8,V_8>}:—
viab(H || S union {<Q_6,V_6>).

viab(H || S union {<Q_9,V_9>} || S union {<Q_8,V_8>}:—
viab(H || S union {<Q_9,V_9>).

viab(H || S union {<Q_8,V_8>} || S union {<Q_{10},V_{10}>}:—
viab(H || S union {<Q_8,V_8>).

Notice that the axioms specify that position Q_8 can be reached following two alternative paths. For example the query

?viab(H1 || {<Q_8,V_8>} || H2)

will be answer enumerating for H1 all the histories leading to Q_8 and giving to H2 the only possible value

{<Q_{10},V_{10}>}

Consider now Fig. 29, which represents a fragment of a SA diagram. The axiomatization could be

viab(H || S union {<Q_1,V_1>, <Q_2,V_2>} || S union {<Q_3,V_3>}:—

viab(H || S union {<Q_1,V_1>, <Q_2,V_2>}), append (V_1,V_2,V_3).

append([],X,X).
append([X|Y],W,[X|Z]):— append(Y,W,Z).

The query

? viab(H || {<Q_3,[a,b,c]>})

will be answered binding H to all possible histories of the kind

H1 || S union {<Q_1,V_1>, <Q_2,V_2>}

and binding V_1 and V_2 to all possible pairs of lists which, when appended, yield [a,b,c].

6. Conclusions

The prototyping strategy of ASPIS is based on executable specification languages based on logic. The executability is obtained via translation into an extended Prolog.

The most interesting aspects of the outlined prototyping strategy are the possibility of checking different aspects of the system and to animate partially specified systems.

In fact, not only queries like *what result do I get if this input is given?* are allowed but also queries of the kind *is there a path which connects this input port to this output port?* are permitted.

The current implementation consists of an analyser and a translator of RSLogic into extended Prolog and a meta Prolog interpreter able to handle sets and constrained clauses.

Acknowledgements

This work would not have been possible without the friendly and cooperative environment of the ASPIS project. Special thanks are due to Guglielmo Casale, Michela Degl'Innocenti, Gianluigi Ferrari and Paola Giovannetti for their contributions to the refinement of the ideas outlined in this paper.

References

ACM 82 ACM SIGSOFT 'Special Issue on Rapid Prototyping' Sw Enging. Notes, 7,5 (1982)

ALA 84 ALAVI, M. 'An Assessment of the Prototyping Approach to Information Systems Development' Comm. of ACM 27,6 (1984), 556–563

ASP 85 The Aspis Project Team 'Domain Knowledge and Protypying Method', ESPRIT Project 401 (1985)

BAL 83 BALZER, B., CHEATHAM, T., GREEN, C. 'Software Technology in the 1990s: Using a New Paradigm' Computer 11,16 (1983) 39–45

BUD 84 BUDDE R. (Editor) 'Approaches to Prototyping' Proc. of Working Conf. on Prototyping, Springer Verlag (1984)

CLA 84 CLARK, K. L., McCABE, F. G. Micro-PROLOG: programming in Logic, Prentice-Hall (1984)

CLO 81 CLOCKSIN, W., MELLISH, C. Programming in Prolog, Springer-Verlag, New York, 1981.

DEA 83 DEARNLEY, P. A., MAYHEW, P. J. 'In Favour of System Prototypes and Their Integration into the Development Cycle' Computer Journal 26,1 (1983), 36–42.

KOW 79 KOWALSKY, R. A. Logic for Problem Solving, Elsevier-North Holland (1979).

RSC 77 ROSS, D. T., SCHOMAN, K. E. 'Structured Analysis for Requirements Definition', IEEE Trans. on Soft. Engin. January 1977.

ROS 77 ROSS, D. T. 'Structured Analysis (SA): A Language for Communicating Ideas' IEEE Trans. on Soft. Engin. January 1977.

SM 185 SMITH, R., KOTIK, G., WESTFOLD, S. 'Research in Knowledge-Based Software Environments at Kestrel Institute' IEEE Trans on Soft. Eng., SE11-11 (1985) 1278–1295.

SOM 82 SOMERVILLE, I. Software Engineering, Addison Wesley (1982).

WAS 82 WASSERMAN, A., SHEWMAKE, D. 'The role of Prototypes in the User Software Engineering (USE) Methodology' in Directions in Human Computer Interaction, Harston H. R. (Ed.) (1984).

ZAV 84 ZAVE, P. 'The Operational vs. Conventional Approach to Software Development' Comm. ACM 27,2 (1984) 104–118.

Lazy software engineering

Drew D. Adams
Laboratoires de Marcoussis, Division Informatique, Centre de Recherches de la C.G.E., Route de Nozay, 91460 Marcoussis, France

Abstract

The applicability of principles and practices of lazy purely declarative programming to various areas of software engineering is outlined. They are shown to help bridge gaps between the different software development stages and activities (specification, design, coding, debugging, testing, documentation) and, consequently, between the different actors involved.

1. Software engineering today

1.1. Review of dataflow ('structured') system analysis
Dataflow system analysis (DSA), as put forth in e.g. [MARC79], differs in three ways from traditional system analysis: (1.1) by its view of what analysis is, (1.2) by how analysis is carried out and (1.3) by the form of its product, the DSA Specification.

1.1.1. Analysis is declarative, logical

1.1.1.1. Analysis is not design
Traditionally, analysis and design have not been clearly differentiated.
The purpose of *analysis* is to logically decompose a system into maximally independent units. It thus describes what a system is or does.
The purpose of *design* is to describe how a system may be physically composed or how it does what it does.

1.1.1.2. Everybody needs to know 'what' the system is
As it describes a system logically, the DSA spec is the principle (human) communication medium and the fundamental system documentation. It is thus

used as the basis of contracts and acceptance tests. As it serves this communication role, it must be simple and understandable to everyone. This is why it's in graphic form.

1.1.1.3. Analysis needs more emphasis

Since the physical realisation of a system ('how') depends on its logical definition ('what'), analysis precedes design and the other realisation activities, logically and temporally. Analysis, i.e. correctly defining the system, is thus the key activity and needs more time and attention than it is traditionally given.

1.1.2. Systems are metabolic; analyse by following the flow of data

Traditionally, analysis being mixed with design, it was more procedural than declarative. System functions (processes) were decomposed by following the flow of control in time, i.e. taking the point of view of the data processor rather than that of the flowing data. On the one hand, such a decomposition is more likely to be in time and space than by level of abstraction: control details are omnipresent. On the other hand, the resulting subsystems are connected by control dependencies as well as data dependencies.

In DSA, system functions are decomposed indirectly, by decomposing their input and output dataflows. This is because data dependencies between processes are minimal dependencies. On the one hand, this means that such dependencies are easier to discover. On the other hand, it means that the result of DSA is a decomposition into maximally independent processes. This in turn lays the basis for a system design which is itself partitioned, with all the advantages of minimal coupling between software and hardware modules. Any additional module connections added during system realisation via synchronisation, shared data etc. are then necessarily explicitly extra, the results of explicit decisions.

1.1.3. The analysis (requirements) specification

The result of analysis is the analysis spec, also called functional spec, requirements spec, and target document. In contract with long-winded traditional analysis specs, a DSA spec is levelled and graphic. It's composed of a *dataflow diagram* (DFD), together with a *data dictionary* (DD) and *primitive process descriptions* (PPDs).

The DFD, also called a bubble chart, portrays the flow of data through the system as named arcs (dataflows) in a directed graph whose nodes are data transforming functions, or processes. Dataflows are defined in the DD using BNF or dictionary-style definitions. Process bubbles are themselves decomposed as lower level DFDs, or are defined via PPDs. PPDs are logic or policy descriptions of any kind: decision trees or tables, structured natural language, pseudo-code, etc.

1.2. Advantages of the DSA approach

The separation of analysis from design as a different activity is essentially the separation of declarative knowledge of what the system does from procedural knowledge of how it does it. As Kowalski [KOWA 84] points out, among other advantages such separation means that declarative knowledge can be 'developed in an incremental fashion, because it's not all tangled together with the way it's used'. It means that both the analysis and subsequent development activities can be simpler and their results more easily modified.

At any given moment during DSA, dataflows and processes may be more or less undefined or informally defined. The final result is hoped to contain only well-defined entities, but there are always some primitive entities which may either be informally described (e.g. 'the seven bit code 0011001') or simply named (e.g. 'a'). The same method thus serves for informal communication as well as formal specification. (Note that 'formal' means neither 'difficult to understand' nor 'without undefined notions'.) The graphic, multidimensional nature of DFDs, together with the removal of dataflow and process definitions from the DFD, has also proven itself as a human communication medium.

1.3. The problem with both traditional analysis and the DSA approach

The main problem with traditional analysis, as well as DSA as originally envisaged, is that their view of the system is static and complete, and they regard the development process as more or less linear: a complete, final analysis constrains a complete, final design which constrains. . . . They are deliberately inflexible in their effort to constrain the implementation to follow the specification. As there is no automatic link between these two (they don't speak the same language) internal rigidity is built into the implementation. Whenever the analysis spec needs changing there are real problems. Sheil [SHEI 84] says it succinctly:

> The redundant descriptions and imposed structure that were so effective in constraining the program to follow the old specification have lost none of their efficacy – they still constrain the program to follow the old specification. And they're difficult to change. The whole point of redundancy is to protect the design from a single unintentional change. But it's equally well protected against a single intentional change.

To a greater and greater degree, as Sheil points out,

- User needs are more or less unknown even to the users. Sometimes they are unknowable before the new system is at least partially built. System definitions shouldn't need to be complete to be useful.
- User needs change and systems evolve. System realisation activities may necessitate redefinition of needs. System definition is thus best viewed as an incremental, ongoing, tentative process. The same is true for other phases of development. Following Sheil, we might speak of 'exploratory' system development.

1.4. Solutions

1.4.1. Incremented development

Although a given definition precedes its realisation, the analysis, design and subsequent development 'phases' might nevertheless proceed in a cascaded or coroutine fashion, partial results of analysis leading to partial designs, etc. This means 'working around' undefined parts of the system.

1.4.2. Non-linear development history management

The changeable nature of definitions and their corresponding realisations means that even such an assembly-line view of things is too linear. Feedback from downstream activities affects those upstream. Can we still speak meaningfully of 'upstream' and 'downstream' then? Yes, in the sense that a given upstream activity determines those downstream but not the reverse. Upstream activities are abstractions of those directly downstream, but of others as well. Analysis is partly determined by, or takes into account, given existing systems or prototypes.

Evolution of a system, its changes, versions etc. may perhaps best be managed by keeping an explicit system development history stream. Such a stream should of course not be limited to being linear but is, more generally, a graph.

1.4.3. From rapid prototyping to executable incomplete specs

Quickly realising a working, if reduced, system based on initial user input can help further elucidate his needs and allow proposed system changes to be explored. The extreme form of such rapid prototyping is an executable analysis spec.

In general, a system definition will be partial at any given time, thus requiring 'working around' the undefined parts, as mentioned above. This avoidance is automatic if execution or evaluation of the spec tolerates undefined terms without complaining. Names can then be given to unknown or partially known entities and definitional details filled in later as they arise, while the spec remains executable. Then both of the weaknesses of the DSA approach, inability to cope with evolution and incomplete definition, are removed.

Being executable means that, although the spec itself only expresses declarative knowledge, there is a behind the scenes procedural interpretation of that knowledge. In fact, as logic programming and functional programming have recently shown us, all system development activities, from abstract system analysis and design through concrete coding and testing, have a declarative, logical interpretation as well as a procedural one.

1.4.4. A single development language

If analysis, design and programming products are constructed according to the same declarative principles, the difference between them is the level of abstraction. If a language permits abstraction and is not limited in its expressiveness to concerns of a certain domain, it might thus be useful in all phases of system development. This, in turn, would remove the need to add rigidity in order to constrain an implementation to follow its spec.

This is not to underestimate the different problems, concerns and expertise involved in the different system development activities; it is to recognize that they share the same logical basis and overall aim: development of modular systems, organised according to logical, or data, dependencies.

1.4.5. Declarative transformations and procedural optimisations

A single system interpretation may not be sufficient for all development purposes. Design and implementation considerations may require transforming the system definition in a way which preserves its declarative interpretation, while providing a different procedural one. Another possibility is to change not the text but the procedural interpreter, in such a way that the declarative meaning is not affected. Such transformations, optimisations, etc. are computer science techniques which may be independent of the particular application and its meaning. Declarative application-specific knowledge is the user's domain.

2. Developing systems declaratively

2.1. The declarative view: where does it come from?

Historically, the declarative view of things reached the world of practical computer science first through DSA, although the theory of computation, coming from the fields of math and logic, has been expressed in declarative form (λ-calculus, combinatory logic) since its origins. System analysts had the luxury to ignore how and in what order data is transformed and to concentrate on describing what changes take place. The only order that counts then is that imposed by data dependencies.

The history of software engineering has essentially been a move from a procedural to a declarative outlook. The former, being an ideological reflection of conventional architecture, has been called the 'intellectual von Neumann bottleneck' [BACK 78]. It is the individual point of view of an active data processor performing sequential actions on passive data. It is inherently imperative.

The declarative view is that of the interdependent dataflows making up the entire system. Time sequences of commands give way to timeless sequences of data. Or, rather, we may regard dataflows as static sequences because we view them in abstraction from processor time. On the other hand, we aren't obliged to consider them passive, static or timeless. It's sometimes worthwhile to regard them as active, flowing in time, where timing is independent of processor time and is simply a question of logical dependency. Such a view is none the less declarative. According to this view, there are no absolute actors, no passive data. There is only reciprocal action by active flowing data, constantly transforming itself. Systems are composed, at all levels, of related dataflows.

Historically, then, the first major effort to 'control control' was the movement referred to as 'structured programming'. Upon its banner it inscribed 'no *GOTOs*', symbolising a mistrust of indiscriminate commandment. The imperative content of '*GOTO*' was relegated to more manageable forms.

The abandonment of procedure occurring now has as its watchword 'no assignment'. In the broadest sense this means 'no commandment', or 'no side effects'. Procedure, time, history, local state are to be represented explicitly, decoupled from processor time and state.

Of course, even the luxury of rising above an individual processor viewpoint was really forced upon analysts by material changes in the nature of computer systems. First, inside the computer, the CPU/centralised memory dualism, based on their relative difference in cost, gave way to other architectural possibilities based on using the same technology for both.

Second, the environment surrounding the computer was transformed to an interactive communications network with heterogeneous elements. And to top it off, all the elements, computers, communication media and peripheral components, evolved at breakneck speed. Abstraction to a dataflow network system view was obligatory.

2.2. Executable declaration: logic programming and (purely) functional programming

Declarative programming has its roots in math and logic. Functional programming (FP) uses as a computation scheme the fact than an expression may always be replaced by its definition, since they mean the same thing. It's thus based on a reduction semantics, whereby an input expression is evaluated, or reduced, to a simpler form having the same meaning.

Logic programming (LP) uses a set of predicate definitions as axioms and theorems and as a computation scheme attempts to prove that a given query expression follows from the axioms and theorems. Along the way, undefined variables are given plausible values and, if a proof is achieved, those values are output as results of the computation. LP is thus relational; arguments of the defined predicates may be taken as input if instantiated, or output if not. The advantage of LP is thus that it is bi-directional. The advantage of FP is that functional expressions are not limited to logical propositions, and that functions may be higher order, i.e. may have functional arguments or return functional results. Hence the current efforts to integrate FP and LP in order to reap the benefits of both. (See e.g. [DERS 85, FRIB 85, YOU 86, REDD 85, ROBI 85, DARL 85].) The arguments in favour of declarative, as opposed to procedural, languages are reviewed in an appendix.

2.3. FP or LP to execute DSA specs

As declarative languages, either FP or LP could be used as the computation scheme for executing a DSA spec. They are both amenable to source transformation and other optimisations, so that they could conceivably be used throughout system development. Kowalski [KOWA 84] makes the point for LP or, rather, he proposes using LP instead of a DSA spec. He claims that a DFD implies a single direction of procedural execution, whereas the declarative knowledge underlying it is indifferent to direction and an equivalent logic program may be executed in either direction. But the DFD, just as the logic program, implies no particular computation whatsoever. The directions of the directed arcs in the DFD, just as the sides of LP or FP definition, indicate only logical (data, or definitional) dependencies, not order of execution. Execution may certainly 'work backwards' from an input dataflow which is defined in terms of another. This doesn't mean we need consider it as an output dataflow.

2.4. A rich system development environment

Adding FP/LP execution to the DSA approach means we might have a single system development environment used by analysts, designers, etc., much as modern CAD/CAM systems span the product development stream for machined parts, etc. Different views of the system would be possible: graphic DFD-style views of processes abstracting from details of subprocess (function) and dataflow definitions as well as linear code-style views, and DD-style views of dataflows. PPDs would be unnecessary as separate, procedural structured English descriptions – all process definitions would be functional. The different views of the system would be so many different editing or definition modes. In addition to dictionary-style constructive data definitions, we might incorporate abstract data type (ADT) definitions a la algebraic specification, in order to gain the essential advantages of object-oriented programming without a procedural concern for object local state. (Cf. [TRRN 85] for an example of inclusion of ADTs in a functional language.) Dataflows might just as well contain functions as first class elements too. We should like access to a knowledge base of equivalence-preserving source transformations and other optimisation techniques, as well as some in-built expertise in recognising and advisory where they might fruitfully be applied. Of particular value would be be a capability of managing the consistency of the system throughout its definition and evolution. De Kleer's assumption-based truth maintenance system (ATMS) is an appropriate model for such a capability [KLEE 86].

2.5. Abstract data type specification

Data, or type, abstraction might be viewed as a competitor to the procedural abstraction of FP/LP as a system organising principle. In the latter, functions are defined individually, according to their arguments. The former defines a data type in terms of certain functions which may have arguments or return values of that type, thus indirectly defining a group of related functions at once. It has a useful role to play, especially where a data type means something to us intuitively, or where we have a particular representation of it in mind. Thus, although the whole point of an ADT is to allow us to abstract from the details of a given representation, the abstraction is nevertheless derived from one or more representations already considered to be useful. As Guttag says,

> The decision about what operations to include within a data abstraction depends upon which operations have implementations that can make effective use of this knowledge [GUTT 83].

Other criteria for packaging functions together as a data abstraction, such as consistency and completeness, are merely formal, ensuring compatibility among component function definitions.

If deciding what gets packaged together is based on implementation, rather than logical, considerations, it should be clear that data abstraction as an organizing principle is mainly useful during design and subsequent phases and that, even then, contrary to what Guttag says, procedural abstraction, not data abstraction, should more often 'provide the primary organisational tool in the programming process' [GUTT 83]. Data abstractions are not defined just to be able to create other data abstractions. (Some who are very object-oriented may not agree!) We normally do things with them – interrelate them in new ways, not each of which is worth regarding as a new data type. E.g., once having defined the integers we may use them to define rational numbers, but we may also define new functions having integer arguments or returning integer values, functions which may never be used as basic operations defining a data type. It would seem to be stretching things to expect all or even most functions to be defined as basic operations of some ADTs.

3. Laziness

3.1. Eager and lazy interpretation

Our declarative view, consisting of active interdependent streams of flowing data, in no way limits such dataflows to finite sequences. If a corresponding procedural interpretation is to reflect this, and if we abandon the point of view of an outside active agent processing passive data, control must be implicit in the interpretation of an expression representing a dataflow. Data, or logical, dependencies may be interpreted as either supply-driven (eager) or demand-driven (lazy) production, but other interpretations are possible as well [AMAM 84, MOLL 83].

Assuming that dataflow production (transformation) takes time, the lazy approach avoids wasting time on production that will never be consumed, whereas the eager approach avoids wasting time waiting for production that could have been consumed right away. In any case, independent flows may be realized in parallel. If parallel processing is only simulated, the eager approach risks starving processes by eternally realising an infinite dataflow. Much current research effort is devoted to determining which processing can safely be done eagerly, otherwise proceeding lazily [CLAC 85, HUDA 86, HUET 79, THAT 84]. That is, interpretation is to be demand-driven, but an attempt is made to foresee necessary demand.

3.2. Laziness as a general principle

Aside from such lazy evaluation [FRIE 76, HEND 76], we might view other attempts to avoid unnecessary activity, or to delay activity not known to be necessary at the moment, as forms of laziness. Laziness then means keeping our options open. From this point of view, software engineering has basically been a movement to promote laziness. We don't always consciously ask ourselves 'isn't there a lazier approach?', but we might be better off if we did. What's often happened is that we've been too lazy thinking about laziness. Or, rather, we've been too eager to promote a given 'solution' without checking out the consequences, i.e. what else depends on it. Of course, hindsight is clearer than foresight.

Some examples of laziness in software engineering:

- *late binding (procrastination)*

The extreme form of this permits unbound expressions. This underlines the distinction between declarative and procedural interpretation. A given expression may, at the same time, mean something to us and yet be undefined to the interpreter. A 'lazy' interpreter, in the general sense, just leaves it alone, without complaining. (All reduction systems are lazy in this sense). Note that this provides automatic 'stubbing' during definition, debugging and testing, and may even be used as a form of in-line documentation.

- *polymorphism*

No sense being overrestrictive in typing. The extreme form here allows untyped expressions and considers typing as a revocable assumption of definition completeness.

- *separation of declarative and procedural interpretations*

Here, the idea is to not impose a particular procedural interpretation on the source text, thus (1) leaving open the possibility of different interpretations and (2) allowing source transformations, both respecting the declarative semantics. Note that a parallel interpretation is a special case of (1). Explicit parallelism via synchronisation etc. is, in general, a less lazy approach.

- *knowledge bases as separate from inference engines*

This is a special case of the separation of declarative and procedural aspects.

- *avoiding redundancy*

This takes many forms, including using centralised data bases and using the same language for different development phases.

• *avoiding unnecessary recalculation*
One example is so-called 'fully lazy' graph reduction, whereby identical subexpressions are only reduced once. Another is the use of a development history management system, allowing previous states to be re-actualised without recalculation. Still another is memorisation. In fact, caching, in general, is a form of recalculation avoidance.

Obviously the list goes on. What developments in computer science aren't lazy, then? Optimisations that make irrevocable assumptions. E.g. state evolution is usually realised via assignment. In and of itself, this limits flexibility. Unless such action is surrounded by compensating management, previous states can't be reinstated, it's difficult to reason about the evolution, etc. Of course, flexibility is relative. Providing explicit state history may be more work (for both the user and the implementation) than it's worth for a given application. A flexible environment gives us the choice whether or not to retain flexibility!

3.3. Managing definition

If the declarative import of a system description is a set of interrelated definitions, and if we want to leave such a set, as well as the individual definitions in it, open to evolution, then we need a procedural interpretation of defining as a dynamic, interactive, incremental activity.

3.3.1.1. Totally undefined

First, consider undefined expressions. There are, in any non-trivial interpretation, some expressions which are totally undefined, in the sense that, unless able to recognise them beforehand, the interpreter will go off on a wild goose chase trying to interpret them. Being totally undefined means that it is not merely a question of non-termination. While interpreting, no progress whatsoever is made; it's not even possible to realise a partial result. Barendregt [BARE 75] has called lambda expressions with this property 'unsolvable'. Such expressions may semantically be considered to denote 'bottom'. They are undefined for us, but the interpreter is content to keep interpreting them. We won't concern ourselves further with such expressions here.

3.3.1.2. Not yet defined

There are also expressions which we may regard as meaningful but which have not been defined for the interpreter. These fall in two classes. First, there are expressions about which we assume nothing regarding their values or interpretations. Perhaps they'll be defined later, and we can't know ahead of time how they should be treated. Or perhaps we just want to use them to represent unknown values. E.g. (*Appended* x [2]) might represent an unknown list with last element 2, if x is undefined. Likewise, ($>$ x 3) might represent an unknown boolean value, i.e. a constraint on x.

3.3.1.3. Self-defined

Second, there are undefined expressions, called constructors, about which we assume only that their syntactic distinctness determines semantic distinctness. They are thus distinct primitive, or canonical, expressions, and might be regarded as being defined as themselves. How an interpreter knows (or even decides!) which undefined expressions are to be considered constructions and which not, we won't consider, except to note that a lazy, flexible approach would leave the question open, allowing a given method (e.g. a syntactic difference such as upper/lower case) to be temporarily assumed, etc.

The importance of constructions lies in the treatment of equality. As syntactically distinct constructions are semantically distinct, we may consider them unequal. E.g., if expressions e_1 and e_2 are distinct constructions, then (= e_1 e_2) may be reduced to (equivalently replaced by) the construction *False*. If at least one of e_1, e_2 is undefined but not a construction, no such reduction to *False* is possible.

3.3.2. Definitions as assumptions

A set of definitions may be regarded as a possible world, or set of assumptions. Such a set may be managed via an ATMS, i.e. whenever the set is changed, repercussions are taken into account. Likewise, this permits individual definitions, or the set as a whole, to temporarily be assumed complete or not. Examples of completeness assumptions are the 'negation as failure' [CLAR 78] inference rule or the 'closed world hypothesis' [REIT 78] and treatment of undefined, non-construction expressions as errors – in fact, defaulting, in general. Incompleteness assumptions are used to represent unknown or partially known values.

3.3.3. Most definition is monotonic

Regarding a set of definitions as an assumption base, we may consider two kinds of changes to it. Some changes are merely independent additions; some are non-monotonic alterations. Definition, just as knowledge acquisition, is mostly a monotonic operation. Of course, there develop inconsistencies and there are about-faces, but in between such 'reclassifications' there is usually an awful lot of accumulation of independent material. After checking and determining that a new definition is independent of those already present, adding it in no way affects previous results, excepting of course those based on assuming the definition set was complete.

3.3.4. Monotonic definition preserves referential transparency

Disregarding such assumptions, it's clear that the property of referential transparency (RT) is preserved when adding independent definitions. This property states that the same expression, in the same definitional context, always denotes the same value. If the operational semantics is, e.g., reduction, then an expression is 'evaluated' by replacing it with an equivalent one, in the sense that they both have the same denotation. Equivalence may be operationally determined by the functor =. Then, if (= e_1 e_2) may be reduced no further, so that it represents an unknown boolean value, we know that e_1 and e_2 aren't both constructions. If one or both are subsequently defined in such a way that the equation then reduces to *True* or *False*, we may consider that the equation has not changed value but, rather, its previously unknown boolean value has become known. Yet, the result of reducing the equation is different before, from after, the new definition. It is not syntactically identical, but it is equivalent, as determined by =.

Note that, in order to guarantee RT, it is not enough that the expression to be defined be undefined; it must also not be a construction. For, if *c* and *k* are different constructors, then (= *c* *k*) reduces to *False*. If *c* is then defined as *k*, the same equation reduces to *True*. In checking for independence of a new definition, then, it is important that constructions be considered to be defined as themselves.

3.3.5. Lazy definition

Regardless of how assumptions are managed, then, it's desirable to have available definition sets which are not inconsistent, preserving RT with all its advantages. Provided definitions are kept open, even the simplest reduction implementation with only a simple mechanism for ensuring definition independence (such as non-unification of reduction rule left hand sides), may thus be assured of RT. On the other hand, the eager approach of assuming all definitions to be closed (e.g. assuming undefined expressions to be errors) requires sophisticated assumption management right from the beginning, in order not to sacrifice RT. That is, every change to the definition set is then non-monotonic. Laziness means keeping our definitions, as options, open. It should be obvious that type-checking is a form of definition closure, as well. The point is not to outlaw error processing etc., but to avoid or delay implementing it, using it expressly when it's useful, and recording the decision to do so as an explicit, revocable development decision.

4. Conclusion

We've argued that having both a declarative and (one or more equivalent) procedural interpretations of a system, but keeping them separate, is an idea which could fruitfully be borrowed from the world of artificial intelligence and applied to software engineering. The importance has been stressed of using a simple declarative language which tolerates undefined expressions, throughout system development. Development then proceeds as an evolutionary stream (graph) of logical changes to the system definition. Monotonic changes to the procedural interpretation are carried out either via source transformation or by changing the interpretation mechanism but, in either case, in such a way as to preserve the declarative semantics. The design decisions and analysis redefinition which punctuate system development, and which produce non-monotonic changes in the declarative semantics, are best handled as revocable assumptions.

APPENDIX: Advantages of lazy declarative languages

Here is a brief summary (cf. [BACK 78]) of the advantages of lazy declarative languages over eager procedural languages:

- *indefinite*

Indefinite and non-terminating entities may be explicitly depicted and manipulated as first class objects. The values denotable by non-lazy languages are: completely unknown (bottom), exactly known, and contradictory (top). Lazy languages allow all combinations of partial knowledge.

- *controllable*

Explicit declarative control, time, state, history are user manipulable and independent of machine time, state, etc. 'Control' of 'combinatory explosion' becomes a matter of describing search paths and spaces, deduction and proof methods etc. as (possibly infinite) data structures.

- *referentially transparent and side-effect free*

Equal expressions are always interchangeable, everywhere. What you see in the program text is what you get, every time.

- *simple/concise*

Programs are small since details of computation are abstracted from. In a procedural language, the need to follow control flow plunges one immediately into details. In a declarative language, processes are represented as logical data transformations, not as physical machine events.

- *semantically clean*

First, program operational semantics need not take into account CPU states and their transitions as in procedural languages, since there are no side-effects. Second, program execution is expression simplification which preserves meaning (reduction); i.e. referential transparency means that expression denotion is constant throughout simplification.*
Third, denotable value domains are rich: denotations of expressions may approximate each other according to their information content (see indefinite).

- *conceptually clear*

Programmers are able to separate the conceptual domain of the application from that of computer science per se. With respect to the former, what is described is what the system does, not how. Computation may be incremental while our thinking may be in terms of whole complex dataflows.

- *higher order*

Functions of all orders may be first class language elements.

- *correct/reasonable*

Proofs of correctness, reasoning may be based on static analysis of program text – the dynamic computational process in the computer need not be considered. A simple substitution model of computation is sufficient.

- *transformable*

Programs and data text are easily manipulated for 1) efficiency improvement, and 2) program modification and generation.

- *modular*

Data dependencies between program modules are necessary and sufficient, hence minimal. All other connections (flags, synchronisation, etc.) are superfluous from a logical point of view.

- *inherently parallel, asynchronous*

Data dependencies impose certain sequencing; otherwise, computation order is up for grabs.

* This is in contrast with e.g. LISP, where there can be no simple correspondence between expression and denotation; the notion of 'evaluation' confounds denotation with reduction. For example, the expression (*quote* (*foo x*)) may be said to denote the expression (value) (*foo x*), to which it evaluates, but it may not be replaced by that expression in all contexts. If *foo* is defined as follows:

(*defun foo* (*x*) (*quote* (*2 3*)))

Consider:

(*car* (*quote* (*foo 5*)))

which evaluates to the expression *foo*, which denotes the function *foo*, whereas

(*car* (*foo 5*))

evaluates to the expression (numeral) *2* which denotes the number *2*.
The expression (* *5 4*), on the other hand, may be said to reduce to the expression *20* to which it evaluates, as both expressions denote the same number *20*.

References

BACK 78 BACKUS, JOHN, 'Can Programming Be Liberated from the von Neumann Style? A Functional Style and Its Algebra of Programs' *Communications of the ACM* **21** (8) pp. 613–641 (1978).

BARE 75 BARENDREGT, HENK P., 'Solvability in Lambda Calculi,' pp. 209–219 in *Colloque International de Logique, Clermont-Ferrand, France,* CNRS, Paris, France (1975).

CLAC 85 CLACK, CHRIS and SIMON L. PEYTON-JONES, 'Strictness Analysis – A Practical Approach,' pp. 35–49 in *Functional Programming Languages and Computer Architecture, Lecture Notes in Computer Science,* ed. J.-P. JOUANNAUD, Springer-Verlag, Nancy, FRANCE (September 16–19, 1985).

CLAR 78 CLARK, KEITH L., 'Negation as Failure,' pp. 293–322 in *Logic and Data Bases,* ed. HERVE GALLAIRE & JACK MINKER, Plenum Press, New York (1978).

DARL 85 DARLINGTON, J., A. J. FIELD, and H. PULL, 'The Unification of Functional and Logic Languages.' in *Logic Programming, Functions, Relations and Equations,* ed. D. DEGROOT & G. LINDSTROM, Prentice-Hall, Englewood Cliffs, New Jersey (1985).

DERS 85 DERSHOWITZ, NACHUM and DAVID A. PLAISTED, 'Logic Programming Cum Applicative Programming' *Symposium on Logic Programming*, pp. 54–66 IEEE, (July 15–18, 1985).

FRIB 85 FRIBOURG, LAURENT, 'SLOG: A Logic Programming Language Interpreter Based On Clausal Superposition and Rewriting' *Symposium on Logic Programming*, pp. 172–184 IEEE, (July 15–18, 1985).

FRIE 76 FRIEDMAN, DANIEL P. and DAVID S. WISE, 'CONS Should Not Evaluate its Arguments' *Automata Languages and Programming*: *Third Int'l Coll.*, pp. 257–284 Edinburgh University Press, (July 1976).

GUTT 83 GUTTAG, JOHN, 'Notes on Using Type Abstractions in Functional Programming,' pp. 103–128 in *Functional Programming and its Applications*, ed. J. DARLINGTON, P. HENDERSON, D. A. TURNER, Cambridge University Press (1983).

HEND 76 HENDERSON, PETER and JAMES H. MORRIS, Jr., 'A Lazy Evaluator' *Third Annual ACM Symposium on Principles of Programming Languages,* pp. 95–103 ACM, (January 1976).

HUDA 86 HUDAK, PAUL and JONATHAN YOUNG, 'Higher-Order Strictness Analysis in Untyped Lambda Calculus' *Thirteenth Annual ACM Symposium on Principles of Programming Languages*, pp. 97–109 ACM, (January 13–15, 1986).

HUET 79 HUET, GERARD P. and JEAN-JACQUES LEVY, 'Call By Need Computations in Nonambiguous Linear Term Rewriting Systems,' Tech. Rept. 359, INRIA, Le Chesnay, FRANCE (1979).

KLEE 86 KLEER, JOHAN, DE, 'An Assumption-Based TMS' *Artificial Intelligence* **28** (2) pp. 127–162 North-Holland, (March 1986).

KOWA 84 KOWAKSKI, ROBERT, 'Software Engineering and Artificial Intelligence in New Generation Computing' *Future Generations Computer Systems* **1** (1) pp. 39–49 North-Holland, (July 1984).

MARC 79 MARCO, TOM DE, *Structured Analysis and System Specification,* Yourdon, New York (1979).

MOLL 83 MOLLER, BERNHARD, 'An Algebraic Semantics for Busy (Data-Driven) and Lazy (Demand-Driven) Evaluation and Its Application to a Functional Language,' pp. 513–526 in *Automata, Languages and Programming, Lecture Notes in Computer Science,* ed. J. DIAZ, Springer-Verlag, Berlin (1983).

REDD 85 REDDY, UDAY S., 'Narrowing as the Operational Semantics of Functional Languages' *1985 Symp. on Logic Programming*, pp. 138–151 IEEE, (July 15–18 1985).

REIT 78 REITER, RAYMOND, 'On Closed World Data Bases,' pp. 55-76 in *Logic and Data Bases*, ed. HERVE GALLAIRE & JACK MINKER, Plenum Press, New York (1978).

ROBI 85 ROBINSON, J. ALAN, 'Beyond LOGLISP: Combining Functional and Relational Programming in a Reduction Setting,' Syracuse University Research Report, Syracuse University, Syracuse, New York (April 1985).

SHEI 84 SHEIL, B. A., 'Power Tools for Programmers,' pp. 19–30 in *Interactive Programming Environments,* ed. DAVID R. BARSTOW, HOWARD E. SHROBE & ERIK SANDEWALL, McGraw-Hill, New York (1984).

THAT 84 THATTE, SATISH, R., 'Demand Driven Evaluation With Equations', Tech. Rept. CRL-TR-34-84, Univ. Michigan, Ann Arbor, Michigan (1984)

YOU 86 YOU, JIA-HUAI and P. A. SUBRAHMANYAM, 'Equational Logic Programming: An Extension to Equational Programming' *Thirteenth ACM Symposium on Principles of Programming Languages,* pp. 209–218 ACM, (January 13–15, 1986).

AMAM 84 AMAMIYA, MAKOTO and RYUZO HASEGAWA, 'Dataflow Computing and Eager and Lazy Evaluations' *New Generation Computing* **2** pp. 105–129 Ohmsha, and Springer-Verlag, (1984).

TURN 85 TURNER, DAVID A., 'Miranda: A Non-strict Functional Language with Polymorphic Types,' pp. 1–16 in *Functional Programming Languages and Computer Architecture, Lecture Notes in Computer Science*, ed. J.-P. JOUANNAUD, Springer-Verlag, Nancy, France (September 16–19, 1985).

Unix* – A viable standard for software engineering?

JOHN FAVARO

Introduction

The catchword 'standard' has been used in conjunction with the Unix operating system with increasing frequency in recent times. This situation reflects a growing need among software developers for standards not merely at the programming language level, but at the level of the programming *environment*.

Evidently, Unix has been perceived as the current best hope of achieving that goal. Yet there are significant obstacles to the standardisation of Unix. In this paper we will take a look at recent efforts toward the standardisation of Unix and examine some of the problem areas in detail.

Versions of Unix

Before discussing the standardisation of Unix, we should first consider the motivation that led to the standardisation effort: namely, the many versions of Unix.

Currently, Unix systems fall into three basic categories:

- those systems being marketed by AT&T itself.
- those systems that are Unix-based, but marketed by other companies with a license from AT&T.
- 'Look-alikes': those systems that are marketed *without* a license from AT&T.

AT&T alone markets a number of versions of Unix, including Version 6, PWB, Version 7, 32V, System III and System V.

* Unix is a trademark of AT&T Bell Laboratories.

The bulk of the other Unix versions fall into the second category. They tend to be based upon Version 7 or System III. Systems in this category include XENIX, UNIPLUS, VENIX, IS/3, UTS.

It is into this second category that Berkeley Unix falls. Berkeley Unix was originally based upon 32V, and is now being distributed as 'Berkeley Software Distribution 4.2'.

Some examples of systems in the third category include COHERENT, IDRIS AND UNOS.

This adds up to a staggering number of versions of Unix indeed! Yet this picture is somewhat deceiving; for, only four of these versions have really been taken into serious consideration in the standardisation efforts:

Version 7

Version 7 marks the beginning of 'modern times' for the Unix system. Up to that point, the system had been mainly used in research circles, and had not yet acquired the facilities now considered to be basic to the Unix programming environment, such as the standard I/O library.

Although Version 7 was officially released in 1979, it already existed as early as 1977 when Unix 32V was derived from it. It was the first port of Unix to the VAX, and laid the foundation for Berkeley Unix.

In fact, Version 7 owes much of its significance to the fact that it was the first version to be ported in earnest to the micros. In 1980, ONYX made the first micro port of Version 7, to the Z8000. Others followed suit quickly. In particular, Microsoft ported XENIX to a number of micros.

Version 7 was the last version of Unix that was actually produced by the research group that originally developed Unix. For that reason, the name was changed in later commercial releases.

System III

AT&T released System III in 1981. This was AT&T's first attempt to support Unix officially. A new pricing policy was introduced with System III, which finally provided for binary sublicenses in place of the exhorbitantly expensive source licenses that had previously been required. Technically, System III consolidated the best of Version 6, PWB and Version 7.

But the real significance of System III was the commitment of AT&T. This provided the necessary confidence needed by commercial vendors to base their derivatives on System III, and as a result there are quite a few commercial systems now based on System III.

System V

With the introduction of System V in January 1983, AT&T consolidated its Unix marketing effort. Up until this point, System V had actually been in use *internally* at Bell Labs, but System III was being marketed *externally*. Now this discrepancy no longer existed.

Commercial support was strengthened even further, with the introduction of periodic updates and hotlines to support centers.

Recently System V has been upgraded with virtual memory and file locking facilities.

Berkeley 4.2 BSD

In 1976, Ken Thompson spent a year at the University of California at Berkeley, bringing Unix with him. At this time, a period set in of enhancements so important that it is no longer possible to leave Berkeley out of a thorough discussion of Unix.

These enhancements include:

- The C-Shell. The most important alternative to the Bourne Shell, the C-Shell is preferred by many for interactive use because of facilities for aliasing (renaming commands) and command history lists for recall execution.
- Improved terminal handling. The *curses* and *termcap* packages, as well as the screen editor *vi*.

In 1979, with the 3.0 Berkeley Software Distribution (BSD), virtual memory came to Unix. The large address space paved the way for new applications – for example VAXIMA, the Berkeley implementation of the MACSYMA symbolic and algebraic manipulation system originally developed at MIT.

With the release of 4.2 BSD in October 1983, networking came to Berkeley Unix. Communication facilities based upon the US Department of Defense standard Internet protocols TCP/IP were integrated into the system. Furthermore, the file system was redesigned for higher throughput by taking advantage of larger block sizes and physical disk characteristics. These two additions alone were sufficient to insure Berkeley Unix an important position in the Unix world today.

It is these four Unix variants upon which we will focus our attention in the following discussion; for, taken together, they raise all of the major issues of the current standardisation effort.

The formation of /usr/group

The standardisation problem actually began as early as 1979, when the first ports of Version 7 were undertaken. In seemingly no time at all, many variants sprang up in the commercial world. In recognition of this development, the /usr/group organisation was founded soon afterwards in 1980. The organisation took its membership from the commercial world, yet was vendor-independent.

It did not take long for the members of the organisation to come to a decision about how they wished to spend their time: only a year later, in 1981, the /usr/group Standards Committee was formed.

The Standards Committee included participants from a broad range of vendors. Conspicuously, one of the most enthusiastic and active participants was AT&T. The committee set as its goal a vendor-independent standard for commercial applications.

The 1984 /usr/group proposed standard

The result was presented in 1984 in the form of a proposed standard[1]. The long term goal set for this proposed standard is ANSI and ISO Standardisation, and indeed, an IEEE 'Working Group on Unix Operating System Standards' has since been formed to pursue this goal, using the proposed standard as a base document.

What is contained in the proposed standard? Actually, this question is best approached by asking its complement: What has been *left out* of the proposed standard?

The proposed standard does not specify:

- The user interface (shells, menus, etc.)
- user accounting
- terminal I/O control
- in fact, *most* of the over 200 commands and utilities of the Unix system!

Then, we might ask, what *is* in the proposed standard? The standards committee decided that the best way to achieve portability would be to concentrate on only two areas:

1. System calls

The Unix kernel interfaces with applications programs through a set of system calls. These shield the programs from such internal matters as details of memory management, scheduling, I/O drivers, etc. These calls are described in Chapter 2 of the Unix *Programmer's Reference Manual.* The proposed standard defines a set of 39 system calls that are to be used by all applications programs.

2. The C language libraries

Chapter 3 of the Unix *Programmer's Reference Manual* describes the set of library routines normally available to programs written in the C language. The proposed standard defines a version of this library that is to be used by applications programs.

File locking

In the entire set of system calls specified in the proposed standard, only one *extension* to the set of system calls available on most Unix systems (in 1984, at least) appeared: file and record locking.

This fact testifies to the enormous importance of file and record locking in commercial and data base applications.

The new system call is *lockf(2)*. It may be used for record locking and for semaphores. Using *lockf*, a program may define 'regions' of a file which are locked individually.

Lockf may be used in either an 'advisory' or 'enforced' form. The advisory form may be circumvented by using normal *read* or *write* system calls. Thus, the advisory form assumes that the processes are explicitly cooperating with each other in a 'friendly' way.

The enforced form protects a region even from those who have no knowledge of the facility. The *lockf* definition specifies that deadlock must be guaranteed between processes related by locked resources.

As defined, *lockf* represents a compromise: on the one hand, processes not using locks don't need to know about them; but because of this, deadlocks may occur.

The System V interface definition

We noted earlier that AT&T has been an active participant in the /usr/group standardisation effort. Thus it is not surprising that the proposed standard basically reflects the AT&T world.

It would be quite a feather in AT&T's cap to have an official Unix system standard whose contents correspond to the flavour of Unix marketed by AT&T. The /usr/group proposed standard represents the first step in that direction.

The second step in that direction was taken in January of 1985 with the announcement of the System V Interface Definition[2]. This document defines a minimum set of system calls and library routines that should be common to all operating systems based on System V. If that sounds familiar, it is no coincidence: the document is virtually identical to the /usr/group proposed standard in content. A separate chapter of the document carefully describes those areas in which differences remain, and includes plans for eventual migration towards total compatibility.

AT&T is attempting to back up this document with promises of adherence of future releases of System V to the Interface Definition. These promises have taken the form of two so-called 'Levels of Commitment'. Each component in the interface has a commitment level associated with it. A component with Level 1 will remain in the definition and be migrated in an upwardly compatible way. A component with Level 2 will remain in the definition for at least three years (although it could be dropped later).

The /usr/group proposed standard defines, as discussed before, only a minimum set of functions and ignores the many commands and utilities of Unix. But AT&T was interested in capturing the full functionality of System V in its definition. The solution adopted was to unbundle System V into a 'base' and 'extensions'. The 'base' part corresponds to the /usr/group standard.

The components of the base fall into the following categories:

- Operating system services
- Error conditions
- Signals
- Other library routines
- Header files
- Utilities
- Environmental variables
- Data files
- Directory tree structure
- Special device files

Even within these categories, the components are more finely partitioned. For instance, within the Basic Operating System Services category, the low-level process creation primitives *fork* and *exec* are segregated into a group that should be avoided whenever possible in favour of the more general *system* primitive. Similarly, the use of low-level *read* and *write* operations is discouraged whenever routines from the Standard I/O Library will suffice.

A System V Kernel Extension has also been defined. The functions provided in this set have mostly to do with the semaphores and shared memory of System V. Why wasn't this directly included in the base? We will have more to say about the problems of compatibility at this level later.

The other planned extensions fall into these categories:

- Basic Utilities
- Advanced Utilities
- Software Development
- Network Services
- Large Machine
- Graphics
- Basic Text Processing
- User Interface Services
- Data Base Manager

Verification test software

To tie all of this together, AT&T has reached an agreement with Unisoft Systems for the development of a verification test software package. This test software will determine whether a derivative of System V actually meets the definition. Clearly, AT&T hopes that this validation suite will attain *de facto* the same kind of status in the Unix world that, say, the Ada Validation Suite has in the Ada world for conferring the official stamp of approval on a system.

Fundamental obstacles to standardisation

When we consider the two areas on which the proposed standard concentrates, we find that one of them is relatively unproblematic: the C Language Libraries. The libraries contain a set of stable, well-understood routines, which can be ported to different Unix variants with little trouble. Even the C language itself is about to achieve standardisation – the ANSI X3J11 group is on the verge of defining a standard for C, based upon System III. Thus it will not be a radical departure from the current situation, but rather a codification of the language as it is today.

It is the *other* area addressed by the proposed standard that presents the major obstacles to a successful standardisation attempt: namely, the system calls. For, the system calls reflect to a much greater extent than the C libraries the basic structure of a Unix variant's kernel.

Currently, several distinct groups rely on the different set of system calls available on the major versions:

- Such *de facto* industry standards as XENIX have heretofore been based on the Version 7 kernel. In recognition of this, the proposed standard 'strives for compatibility with Version 7 whenever possible'.

- The proposed standard has based its set of system calls on System III.

- The recently defined *Portable Common Tool Environment*[3] owes much of the nature of its system calls to System V. In addition, the proposed standard will eventually migrate toward System V.

- Finally, the majority of the participants in the world of high-performance networking workstations relies heavily on the kernel facilities provided by Berkeley Unix.

Where do the incompatibilities lie that give rise to these different groups? Although incompatibilities exist in many places, such as differing file system implementations, the problem can best be characterised by the differences in one particular area: *interprocess communication.*

Let us take a closer look now at the facilities provided by each of the four major versions for interprocess communication.

Interprocess communication – Version 7

The following basic set of system calls related to interprocess communication was defined in Version 7:

signal: This defines the response to *pre-defined* external events, such as interrupts, alarms, hangups and hardware errors.
kill: This call is used to send signals to processes.
pause: A process suspends its execution pending receipt of a signal.
wait: A process waits for the termination of a child process.
pipe: The well-known interprocess byte stream.

Now, these facilities are adequate for most time-sharing applications, but not for such applications as networking and data bases. Where are the deficiencies? Basically, the problem here is one of flexibility. Among others, the following cases of inflexibility can be identified:

- *signal* only allows pre-defined signal types. Therefore, no extra information can be conveyed.
- *kill* can only send to single processes or to all processes.
- *pipes* can only be used by *related* processes.

In System III, facilities were introduced to address exactly such problems of inflexibility.

Interprocess communication – System III

Three new facilities were added in System III to handle the problems discussed above:

First, the so-called 'FIFO' file was introduced. The FIFO file is a special file, just like a pipe. But FIFO files have Unix system file names, not just file descriptors (thus they are called 'named pipes'). Furthermore, and more importantly, they can be used by *unrelated processes*, thereby increasing their flexibility enormously.

Secondly, two new, *user-defined* signals were introduced into the set of allowable signals. This allowed extra application-specific information to be conveyed with signals.

Thirdly, *process groups* were introduced. Thus a common basis was established for sending signals. Now the *kill* system call could send signals to all members of a process group.

Yet even these improvements had their problems. Two specific ones may be identified:

- Signals are not a sound basis for interprocess communication and synchronisation. They are not queued, so signals could be lost. The /usr/group proposed standard explicitly discourages the use of signals for synchronisation.

– Pipes, even named ones, are inadequate in crucial ways. For instance, the byte stream of the pipe is simply too low-level. Message boundaries are not preserved, which in many applications can be a problem.

Additionally, pipes are very slow. To see why, consider the flow of data in a pipe: Data starts in the sender's address space. It is then copied into a kernel buffer. Finally, it is copied *again*, from the kernal buffer into the receiver's address space.

In short, the System III facilities cannot be considered to be a sound basis for process synchronisation. With System V, such a basis appeared for the first time.

Interprocess communication – System V

Three major new features were introduced in System V.

The first was *shared memory*. Shared memory is much faster than pipes, because there is no copying of data. Facilities exist to control access to shared memory, as well as synchronised updating by multiple processes.

However, there is one deficiency of shared memory in System V: processes using it must be *related* by a common ancestor.

This restriction does not exist for the second major new feature, *message queues*. Message queues provide a way for unrelated processes to share data. Messages may have types – for example, a process may request all messages of a certain type from a queue.

The final addition in System V was *semaphores*. This well-known facility provided the much-needed, solid foundation for process synchronisation.

At least two of these facilities, however (semaphores and shared memory) are heavily biased towards applications running on a single processor with all processes sharing the same memory space. In order to pursue its own research goals in the area of distributed computing systems, Berkeley introduced an entirely different set of facilities.

Interprocess communication – 4.2 BSD

The facilities for interprocess communication have been enhanced at Berkeley to an extent unmatched in any other Unix version. They represent an entirely different approach to interprocess communication.

The Computer Systems Research Group concluded that, in order to achieve its goal of truly distributed systems, the interprocess communication facilities should be *layered over networking facilities within the kernel itself*. In addition, the facility has been decoupled from the Unix file system and implemented as a completely independent subsystem[4].

In 4.2 BSD the *socket* is the building block for communication. Sockets are named endpoints of communication within a *domain*. Currently, the Unix and DARPA Internet communication domains are supported.

There are basicallly three types of sockets:

- *stream* sockets, which provide a reliable bidirectional flow of data to unrelated processes. In the Unix domain, pipes are available as a special case of sockets.
- *datagram* sockets, which provided facilities similar to those found in Ethernet.
- *raw* sockets, generally intended for use in the development of new communication protocols.

The lowest common denominator

What do these four versions of interprocess communication have in common? Unfortunately, little more than pipes! And the simple byte stream as the lowest common denominator is very low indeed. Defining a set of system calls to handle only such restricted cases is simply not realistic.

We cannot expect developers programming advanced applications to give up the advanced facilities for interprocess communications discussed above and to accept of a standard based upon a more restricted set. The interprocess communication facilities of the various Unix versions must converge much more before a truly representative set of system calls can be defined.

Future convergence?

The convergence of the AT&T Unix world towards System V will continue. Microsoft has reached an agreement with AT&T whereby the next release of XENIX will conform to the System V Interface Definition.

A Berkeley 4.3 distribution will reportedly be available soon – officially 'sometime before 1995' (this statement constitutes a reaction to the delays and disappointments surrounding the date of release of 4.2 BSD). The 4.3 release will, however, consist essentially of features and enhancements that are necessary to stabilise problems that were identified in the current release.

In a major step towards reconciliation of the two major standards, Sun Microsystems and AT&T have agreed to work together to facilitate convergence of System V with Sun's 4.2 BSD-based operating system.

They will attempt to merge the two standards into a single version. Technical representatives are meeting periodically to define a common applications interface. Sun intends to add complete compatibility with the System V Interface Definition, while maintaining the added functionality of 4.2 BSD that was discussed above.

The hopes for a standard Unix environment

Although the problems of standardisation arising from the kernel incompatibilities discussed above are very real, there is a very large and important class of programs that are not affected by them. These are the software engineering tools used in the production of software, such as editors, report generators, filters, testing tools, document production facilities and compilers. Such programs are well served by the proposed standard in its current form. Thus we should expect that a large part of the Unix software engineering environment can indeed be standardised.

However, in advanced areas such as networking, distributed systems, and real-time systems, Berkeley Unix clearly offers superior facilities. Given the current nonconformity of Berkeley Unix to the standards pursued by most of the rest of the Unix world, a standard Unix software engineering environment will depend heavily on the success of efforts to merge these two worlds into a single one.

References

1 Proposed Standard, /usr/group Standards Committees, January 17, 1984. Obtainable from /usr/group, 4655 Old Ironsides Drive, Suite 200, Santa Clara, California 95050, USA

2 AT&T System V Interface Definition, January 1985. Obtainable from AT&T Customer Information Center, Select Code 307–127, 2833 North Franklin Road, Indianapolis, Indiana 46129, USA

3 PCTE: A Basis for a Portable Common Tool Environment, Functional Specifications, First Edition, August 1984, Bull, GEC, ICL, Nixdorf, Olivetti, Siemens

4 LEFFLER, S. J., FABRY, R. S., and JOY, W. N. *A 4.2 BSD Interprocess Communication Primer*, Computer Systems Research Group, Dept. of EE&CS, Univ. of Calif. Berkeley, 1983

The ASPECT project

J. A. Hall
Systems Designers plc

Introduction

ASPECT was the first Alvey-supported software engineering project and is a collaborative venture aimed at prototyping a multi-language, distributed-host, distributed-target Integrated Project Support Environment. The ASPECT team is led by Systems Designers plc (SD) and the other partners are the Universities of Newcastle upon Tyne and of York, GEC Computers Limited, ICL and MARI. Following the Alvey strategy, ASPECT has started by integrating existing tools, notably Perspective from SD, on UNIX; this environment is being developed by distributing it using the Newcastle Connection and then building in the more advanced results of collaborative research and development.

Requirements

To understand the objectives and strategy of ASPECT, it is necessary to consider the requirements for an IPSE: what we expect it to do, beyond what our current tools provide, to improve the software development process. We can identify four areas where an IPSE can advance the state of the art:

(*a*) *It must support the whole software lifecycle*
Whatever one's view of the software lifecycle, it certainly encompasses a number of phases through which the software progresses and, at every phase, a number of different types of activity: planning, managing, carrying out and recording the phase, for example. An IPSE, therefore, must support all these activities for every phase: more importantly, it must *integrate* the various supporting tools so they form a coherent whole.

(*b*) *It must support development methods*
Software development methods can be characterised by: their data model of software development; the (frequently graphical) notation for expressing this model; the rules which govern the application of the model and the procedures for manipulating it. An IPSE must be capable of supporting all these aspects of a method. Because there is no universal method and new methods are continually being introduced, an IPSE must be configurable, to support many methods, and capable of integrating different methods.

(*c*) *It must deliver power to the user*
We need to harness the raw hardware power now available so that both processing power and io bandwidth are more than enough for the user not to be constrained by the system. This implies that an IPSE must be workstation based and have an effective, responsive man–machine interface.

(*d*) *It must support development in the large*
An IPSE must support teams of people working on common projects. At the physical level this implies networking of machines; at the logical level, version and configuration control, concurrency control, access control and task management must be built into the IPSE.

Key objectives

ASPECT, in addressing these requirements, is concentrating on four key objectives.

(*a*) *Integration and openness*
ASPECT is emphasising the development of an infrastructure for tools, because it is by provision of a powerful set of common services to all tools that integration of tools can be achieved. Tools written for single users can immediately be used on large projects when incorporated into ASPECT, for example, because the infrastructure manages all the problems of controlled sharing between users. It is crucial that ASPECT provide these facilities in an open way so that new tools and methods can be incorporated by the user.

(*b*) *Host distribution*
ASPECT is addressed at developers who may be geographically distributed and who work on large projects using a range of machines including personal workstations.

(*c*) *Good man–machine interface*
A major part of the ASPECT research is aimed at providing an architecture in which software engineers can use the power of, for example, bit-mapped graphics and pointing devices effectively.

(*d*) *Target distribution*
ASPECT is oriented towards the development of embedded systems. In particular we are addressing the specification, development and testing of software for distributed target machines.

ASPECT architecture

The ASPECT architecture addresses, in a simple and general way, the key objectives. It is based on a clear separation between tools and kernel, and the provision by the kernel of a powerful set of common services for structuring and storing information, for communicating with users and other tools, and for manipulating remote targets. These functions are made available to tools through the public tool interface (PTI).

To achieve the required openness the PTI is extensible, to support new methods and tools, and configurable, so a project can impose particular methods of working, if required. Furthermore, since the PTI is the only means by which tools use ASPECT services, it necessarily includes within itself the facilities for its own extension and configuration.

One of the most important requirements on ASPECT is that existing tools, written for the host operating system, should be usable within ASPECT. This is made possible by the open tool interface (OTI). The OTI can be thought of as a subset of the PTI which appears to the tool just like the host operating system. ASPECT is hosted on UNIX*, so the OTI makes available to ASPECT a large collection of existing software development tools.

The PTI service fall into four groups:

- information storage
- man–machine interface
- process invocation and communication
- target services.

Each of these services is provided by a layer of software – in the effect of subroutine library – between the tool and the UNIX Kernel. The PTI offers services at a much high level than those of UNIX. To provide the open tool interface, the PTI includes calls appearing to be UNIX system calls, but even these are processed by ASPECT rather than by UNIX so that all tools, including UNIX tools, are fully under the control of ASPECT. Indeed,

* UNIX is a trade mark of AT & T Bell Laboratories.

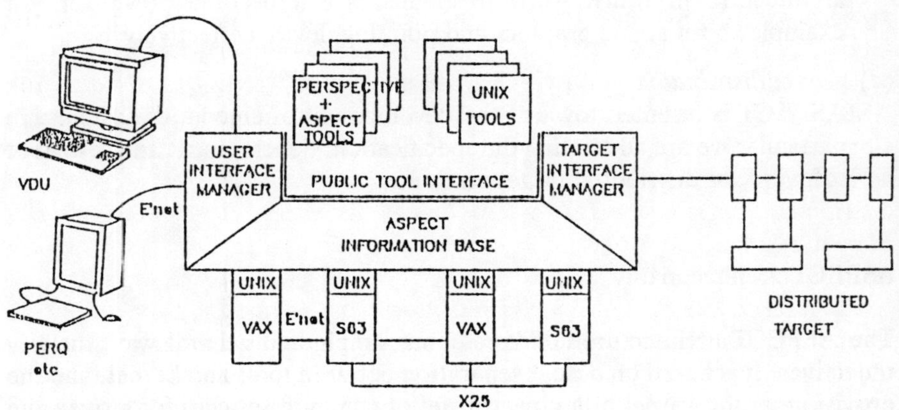

Fig. 32

different open tool interfaces will be provided for different UNIX tools to capture the semantics of the tools' data correctly in the ASPECT information base.

Meeting the objectives

The central component in ASPECT is the information base, and it is the information base which achieves the integration between tools by providing a central, structured repository for all the information they manipulate. The information base is a database but contains in addition:

(*a*) Its own definition. The structural information can be accessed via the PTI in the same way as any other information.

(*b*) Rules. These support not only the integrity constraints of the basic data model, but also user-defined rules which may, for example, be the rules governing the use of a particular development method.

(*c*) Built-in structures to support software engineering. In keeping with the aim of integration, many of the structures (for example version identification) supporting development in the large are provided at the information base level.

The database itself uses a standard architecture, the ANSI/SPARC three level model. This has a conceptual level, with below it an internal level and above it a set of external views, each view presenting the database to a tool in the way required by that tool. The conceptual level uses a standard data model, Codd's extended relational model called RM/T. This is a very powerful combination for providing the required extensibility and openness. The view mechanism not only shields tools from extension to the conceptual model, but it is powerful enough to transform the data to suit almost any tool. In particular the Open Tool Interface is achieved by defining UNIX-like views of the data.

The architecture of ASPECT is implemented on UNIX. In order to run ASPECT on a distributed host we are taking advantage of the Newcastle Connection, a powerful method of linking UNIX systems so that they behave as a single UNIX. ASPECT is building local and wide area networks of distributed UNIX systems. At a level above this, we are developing methods of distributing the information base and, in particular, supporting the logical distribution of the database between separate but interdependent users.

The mmi of ASPECT is, like the information base, aimed at providing a high level of functionality to tools and removing from individual tools concern with the details of user interaction. The mmi architecture has to achieve this across a wide range of users, of devices, of tools, and of interaction styles. At the same time the quality of interaction on powerful workstations is paramount. ASPECT handles this by defining levels of abstraction within the mmi and providing components, with well defined interfaces, to handle these abstractions. This is a major research topic in the project.

In approaching the programming of distributed targets, ASPECT is again looking for general, powerful solutions. We are studying methods for describing target architectures, extending languages to support interprocess communciation on such targets, describing the placement of processes on processors and monitoring the operation of the target.

ASPECT strategy and the Alvey programme

ASPECT is an Alvey second generation IPSE, in that it is clearly based on a database, and is designed for a distributed host. It is not, however, being designed from scratch but is evolving from existing products and ideas. The main starting points were Perspective (a SD environment product), UNIX, and the Newcastle Connection. The initial release of ASPECT is indeed an integration of these components plus an Ada compiler. Meanwhile research has been going on to explore how to move forward form that base.

This research is now being brought together, with our experience from the first release, to define ASPECT – in particular its Public Tool Interface – in some detail. On the basis of this definition, prototypes of ASPECT will be produced and used as vehicles for research and further development. These prototypes will of course reuse as much as possible both of the partners' existing products and of a commercially available DBMS. The results of this work are, in turn, being incorporated into partners' products on UNIX and VAX VMS.

Since much of the ASPECT project is concentrated on the infrastructure, we look to other sources for many of the tools which will run on ASPECT. Some of these tools will be produced by the industrial partners, but ASPECT is also a potential base for tools developed in other Alvey projects and perhaps also other projects like ESPRIT's SPMMS. These tools will use ASPECT most effectively if they exploit the Public Tool Interface, and to aid this ASPECT will produce a formal definition of its PTI using the notation Z, developed at Oxford. Far more tools, of course, have been and will be written simply for UNIX, and ASPECT will integrate these tools through its Open Tool Interface.

Summary

Although ASPECT cannot, of course, address all the problems of software development, it does address the major IPSE requirements and is a prototype of the next generation of IPSEs. In particular it supports the whole lifecycle by providing a sound framework for tool integration; it can be tailored to any method or collection of methods; it provides computing power and a highly functional interface at the disposal of the user and, by its physical distribution and support for controlled sharing it is a powerful environment for development in the large.

Three experimental multimedia work-stations – a realistic utopia for the office of tomorrow

Helmut Balzert
Research Department TRIUMPH–ADLER AG, Nuremberg, West Germany

Abstract

The office of the future needs different multimedia workstations for different user groups. The architecture and the highlights of our multimedia office environment are sketched. The manager workstation has a completely new human-computer-interface: a horizontal flat panel built into an office desk. On the flat panel a touch-sensitive foil is used for input. Virtual keyboards can be displayed on the flat panel if needed by the application. A pencil with a built-in ultrasonic-transmitter is used as a pointing and handwriting device. Our model of multimedia interaction and communication is presented. A detailed explanation of how the processing of office procedures is implemented on our experimental workstations is given.

1. Introduction

In the past, only specialists were able to operate a computer. Generally, a long training phase was necessary: the human had to adapt to the computer. Now software and hardware technology is ready to change the situation completely: the computer is able to adapt to the human.

This ability is a necessary prerequisite for the office of the future. The acceptance of new office systems depends on the following conditions:

- Very short training phases
- Only little change in the current working style or evolutionary change
- Consistent and uniform interface design
- Direct manipulation via alternative multimedia communication channels
- Additional comfort

In the next chapter we will explain our human-computer-communication concept. Some facts about office activities are summarised in chapter 3. The architecture of our multimedia office environment and the hardware highlights are described in the subsequent chapter. Some important office scenarios including our software highlights are sketched in chapter 5. The last chapter contains a resumee and perspectives on the future.

2. Human-computer-communication

In human-computer-communication two forms of communication can be distinguished: explicit and implicit communication [FISC 82].

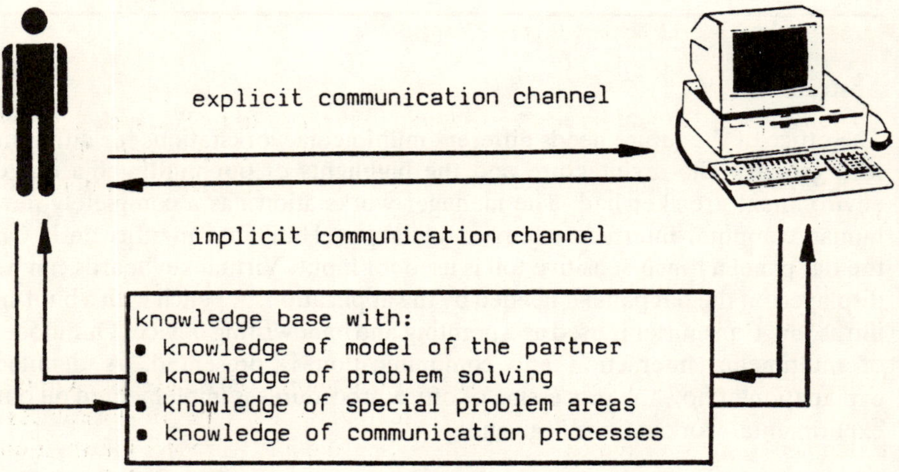

Fig. 33 *Human-computer-communication model*

In order to obtain optimal human-computer-communication the explicit *and* the implicit communication-channel must be very broad.

This article concentrates on the explicit communication channel. Today the explicit communication channel is narrow:

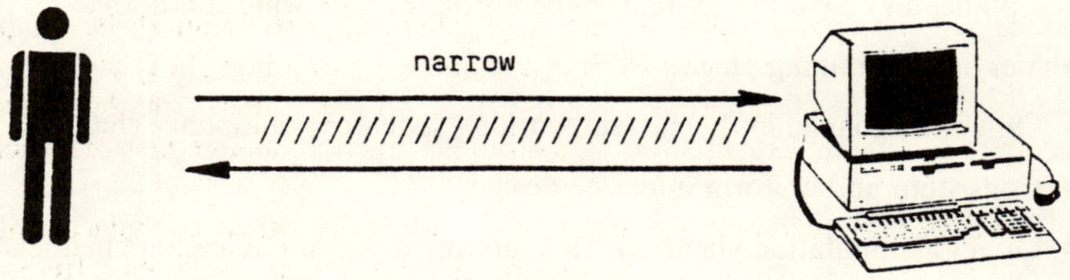

Fig. 34 *Narrow explicit communication channel*

Normally the communication is reduced to input via a keyboard and to output via a display. Modern systems like Xerox Star, Apple Lisa & Macintosh improve the communication using full graphic displays with icons and a mouse as a pointing device.

One way to improve the acceptance of new office systems is the use of a broad multimedia explicit communication channel:

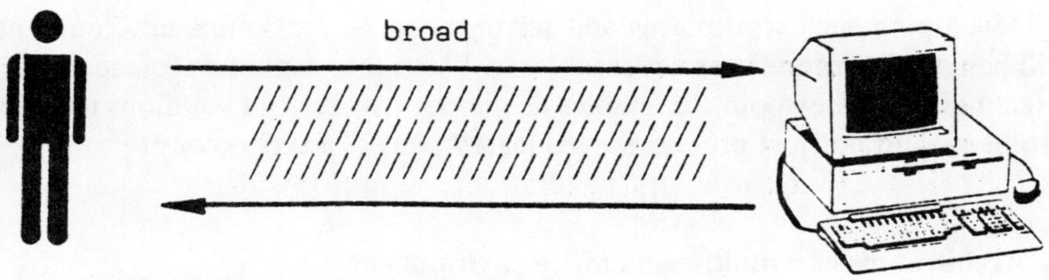

Fig. 35 *Broad explicit communication channel*

Today's software and hardware-technology is ready to realise such a broad communication channel. This basically means that the user had different communication techniques available for the input of commands and information.

3. Office activities

Workstations in offices will, in addition to common applications, support various office activities that are not specific to particular types of workplaces. Examples of such activities are computerised dialling, filtering information from incoming mail, scheduling meetings etc. (see, e.g. [ELNU 80]).

The differences between different workplaces in offices should not be neglected, however. A careful inspection of present-day office work reveals typical differences. Searching, e.g., is more important for knowledge workers than it is for managers. On the other hand, managers will spend more time dictating than clerks or knowledge workers do. Writing will remain a typical activity for secretaries.

Modern information technology certainly will change the office, not only as far as the technology installed is concerned. There will also definitely be a high impact on work structure and work distribution. It would be going too far to address these questions here. It shall be noted, however, that a trend to more flexible structures, to an integration of different functions at a single workplace seems to emerge.

This does not mean that, some day in the future, all office workplaces will look alike. Differences in tasks and attitudes will remain. The three different prototype workplaces that have been developed at TRIUMPH–ADLER's basic research reflect this and have been tailored to three different types of office workplaces:

- a manager's workplace
- a secretary's workplace
- a knowledge worker/clerk's workplace

Developing such workplaces and testing them in real office environments will help to determine user needs and drawbacks that cannot be foreseen from a technological viewpoint. It enables developers to arrive at solutions to office problems and not just present sophisticated pieces of technology.

4. Architecture of a multimedia office environment

Fig. 36 presents the architectural concept of our environment.

Figs 37, 38 and 39 give a visual impression of our existing workstations.

The *hardware highlights* of the three workstations:

The secretary workstation: An ergonomic keyboard with two separated blocks of keys is used for text input. A LCD-display was placed above the free programmable function keys. It displays the current meaning of the function keys, using icons.

A touch-sensitive foil is fixed to the left and to the right of the keyboard. Each touch-sensitive foil is used as a pointing device like a 'static mouse'. We want to ascertain if it is useful to have a pointing device both on the left and on the right side. It is possible to use the foils with different scales: a movement across the foil implies a movement of the cursor across the whole screen with one of the foils. Using the other foil, only a smaller part of the screen is passed with the same movement. Thus, 'global' movements of the cursor over longer distances are carried out easily as well as 'local' movements, i.e. very fine and precise pointing operations. To some extent, this may be regarded as a zooming effect. Substantial results are not available yet as our experiments have just started.

Another advantage of such a foil, in comparison with a mouse, is that it can be used much better as an input medium for graphics and handwriting.

Also integrated in the keyboard is the telephone.

The manager workstation: Because a manager normally does not type a lot of text or information he does not need a traditional keyboard. In addition a lot of managers are not willing to use a keyboard.

We have therefore developed a completely new human-computer-interface: a horizontal flat panel (in our case a plasma display) is built into an office desk.

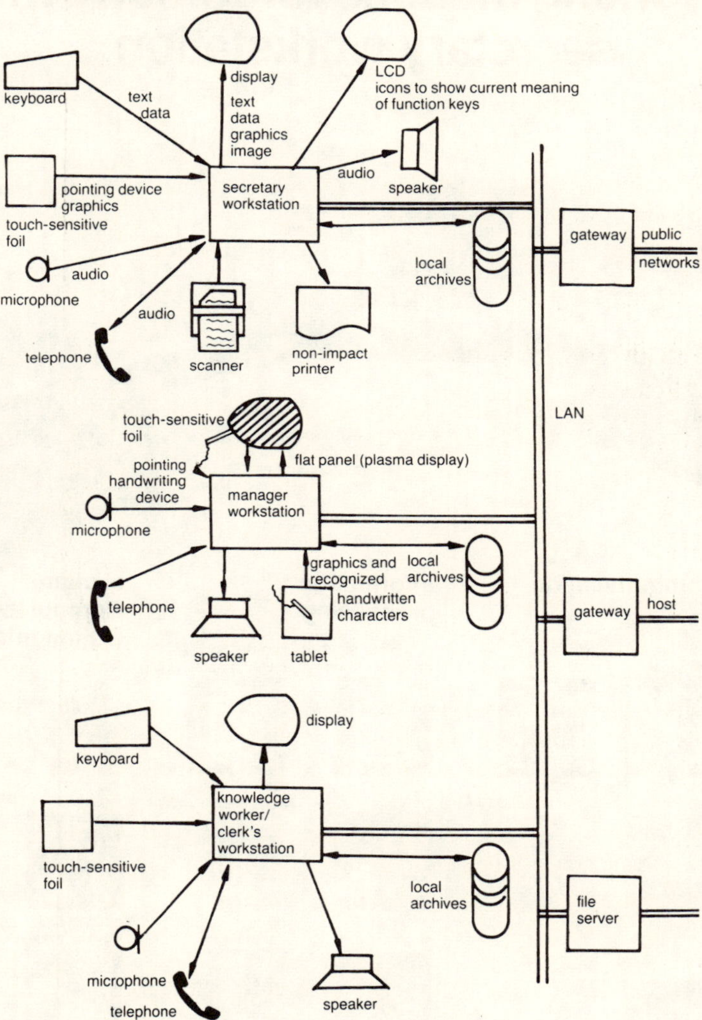

Fig. 36 *Multimedia office environment architecture*

Above the flat panel lies a transparent, touch-sensitive foil, which can be used as an input device for the virtual keyboards that can be displayed on the flat panel, according to the application. The second input medium is a pencil with a built-in ultrasonic transmitter. In one mode it can be used as a a pointing device. Each time the pencil is pressed on the screen, an ultrasonic impulse is transmitted. Two small microphones receive the impulse. A processor computes the position of the pencil with a tolerance of 1/10 mm.

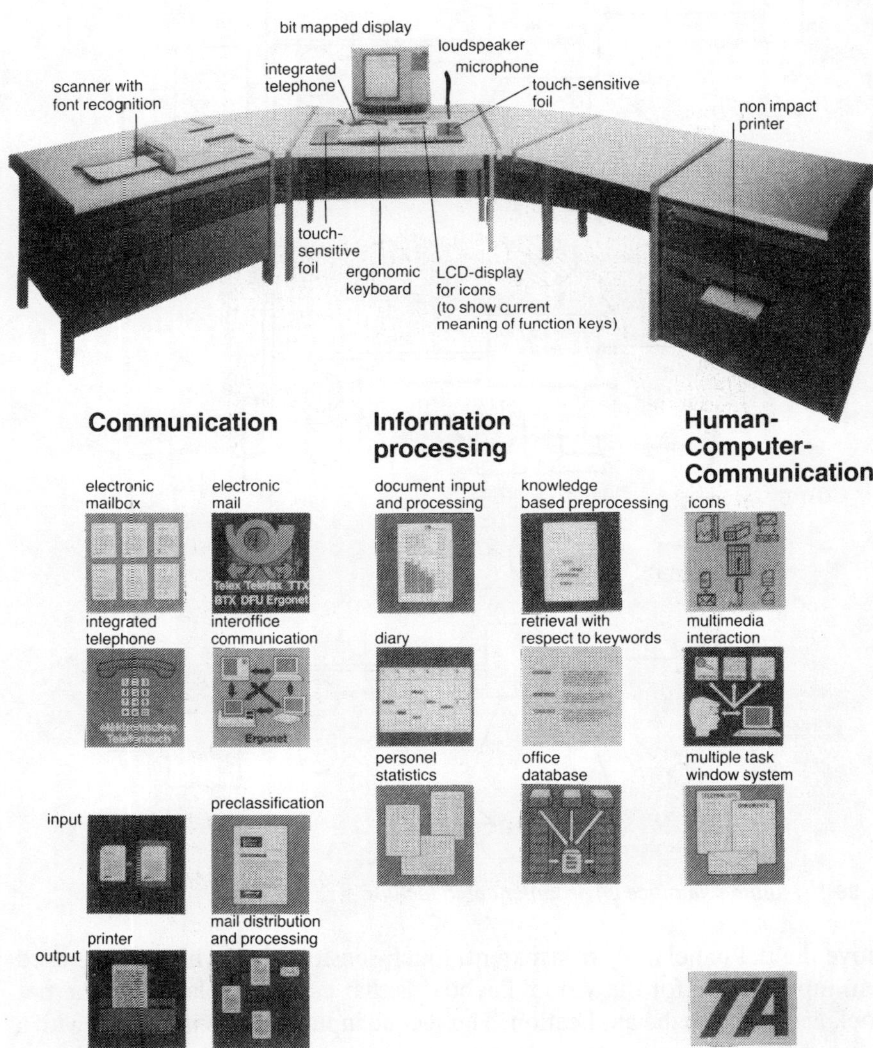

Fig. 37 *The experimental secretary workstation*

Working stations for tomorrow: manager workstation

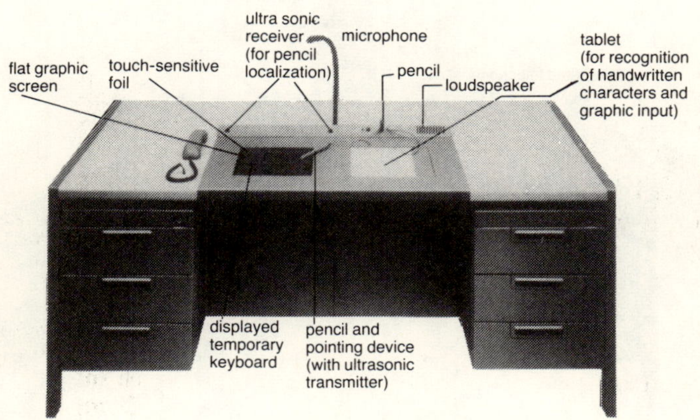

Communication

electronic mailbox

electronic mail

integrated telephone

interoffice communication

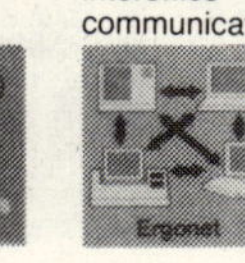

Information processing

document drafting and revision

diary

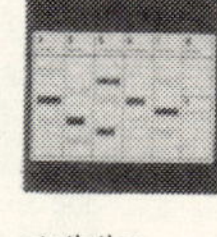

statistics

knowledge based preprocessing

retrieval with respect to keywords

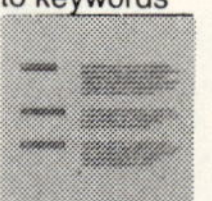

background information

Human-Computer-Communication

multi media interaction

keyboard on request

recognition of handwritten characters

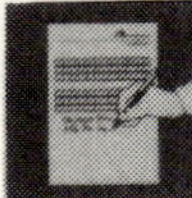

Fig. 38 *The experimental manager workstation*

Working stations for tomorrow: knowledge worker/clerk's workstation

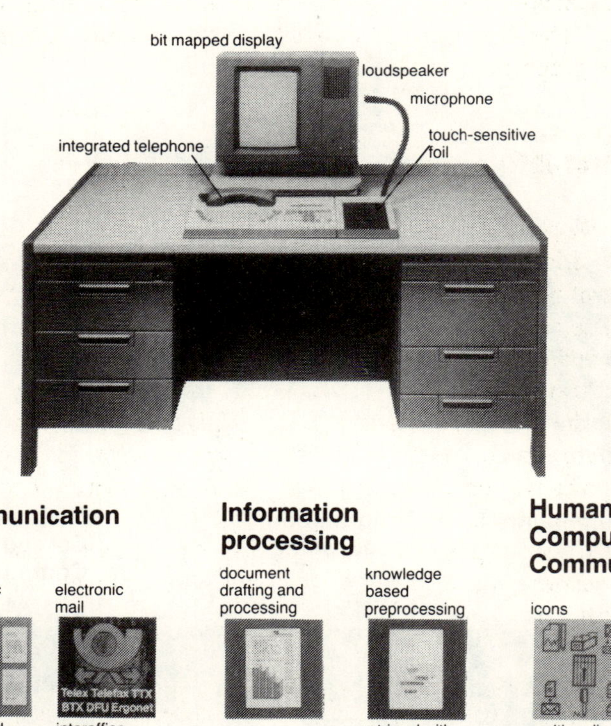

Fig. 39 *The experimental knowledge worker/clerk workstation*

In a stream mode, the transmitter sends more than 100 impulses per second. In this mode, it is possible to do handwriting with the pencil on the flat panel. The handwritten text or graphics will be echoed on the flat panel in real-time.

A graphic tablet in the manager desk allows the recognition of handwritten block letters. It also can be used as an input medium for graphics.

The knowledge worker/clerk workstation: Analogous to the other workstations, this workstation is equipped with an integrated telephone, a microphone and a loudspeaker. The keyboard is connected with a touch-sensitive foil as a pointing and graphics input device.

5. Scenarios of office procedures

Scenario 1: Processing of incoming paper mail
Even in the office of tomorrow not all communication partners will have an electronic mail box. Therefore a part of the incoming mail will be paper mail.

In our office architecture, incoming paper mail will not go further than the secretariate and/or the mail department.

A scanner transforms paper mail into electronic mail. Today we use a text scanner which recognises type-written characters and can differentiate between different type sets. In future, we expect scanners which process mixed modes: graphics will be transferred bit by bit, type-written text and printed text will be recognised.

This transformed paper mail will be processed by an expert system. It tries to locate the sender, the addressee, the date, the subject and type-written signature, if it exists.

If the sender and the addressee can be recognised in the letter, the expert system then searches automatically in the archives, belonging either to workplace, the department or to the whole company, to verify whether the addressee and the sender are known.

If yes, the expert system sends the letter via electronic mail in the addressee's mail box. The secretary will not even notice this process, let alone play a part in it.

In the other case, the secretary gets the letters and documents in her mail box to do a manual preprocessing.

Incoming electronic mail which is not transferred directly to the mail box of the target receiver, is treated in the same way.

Scenario 2: Preclassification, filtering and presentation of incoming mail
Before incoming mail reaches the mail box of the office worker it will be preclassified by an expert system. If the mail did not come via the secretary's workstation to the mail box, then a preprocessing analogue scenario 1 takes place first.

The expert system searches in the personal archives of the office worker to look up all key words which the receiver uses to classify his letters and documents. Then these key words will be looked for in the incoming mail and will be marked (through inverse video).

Each office worker may give priorities to keywords and may also restrict these priorities to a certain duration or specific time intervals. In case the incoming documents contain one or more of these priority key words and, if they exist, obeys the corresponding time restriction, this document will be displayed in a special representation form like inverse video, e.g., or is announced by a loudspeaker, etc.

The opposite is also possible: to determine senders or organisational units whose mail will be rejected. This mail will be sent back to the sender automatically or will be deleted.

This feature is one means of coping with the increasing flood of incoming information that is likely to be a result of widespread electronic information exchange. It will become very easy to send out more and more mail or electronic copies to more and more people. Our approach represents a step towards a flexible and individual solution.

The implementation of our developed expert system is described in [WOEHL 84].

Scenario 3: Processing of mail
If an office worker looks into his mail box, the mail will normally be displayed in the form of so-called miniatures. Miniatures are document icons and show the document considerably reduced in size. Its line and paragraph structure are visible, but not legible, the user gets a small visual image of his mail (Fig. 40).

Fig. 40 *Document miniatures*

This approach helps in choosing documents, since it takes advantage of the excellent visual recognition capabilities of humans. Of course, the user is free to change the presentation of the mail to his needs and personal wishes.

The user chooses a document from his mail that he wants to read by pointing to it. The complete document is then displayed on the screen. The first thing he might do is change the keywords chosen by the pre-classification system. He can then file it, send it to somebody else or start with appropriate tasks. He might, e.g., put some remarks on the letter using handwriting or speech and design an answer, send the letter and the concept of the answer to a colleague of a senior executive.

Scenario 4: Remote access to mail
One of the problems which business people who travel frequently face is the access to their mail. It is not too difficult to develop systems that allow one to listen to speech mail that has come in during the day in the evening via the telephone. In addition to this, the technology of full-synthesis speech will provide the same opportunities for typed mail. The system will read these messages aloud. Of course, the user will be able to give comments on the various messages for his staff.

Scenario 5: Writing and sending letters
Designing a letter traditionally is done on paper or using a dictation machine. New technology allows for new possibilities. Handwriting still is possible. It is not done on paper, however, it takes place directly on the screen, using techniques as described above. This manuscript is then transferred to the secretary electronically. It can be seen in one window of the screen, while the secretary prepares the typed version in a second window. The typed letter is then sent back and signed by the corresponding person, who, of course, again uses handwriting. The letter then is ready for mailing. If the receiver can be reached electronically, this means is used. Otherwise, the text is printed out on a non-impact printer and mailed.

Letters for in-house use that do not have to be type can be sent even more conveniently. The writer of the letter specifies the intended receiver and, if necessary, his address in block letters. Thus both the text of the message and intended addresses are handwritten. The underlying system is able to recognise the information on the addressee. A mere push on a 'mail' key then is enough to transmit the mail to the intended receiver.

Scenario 6: Using the telephone
When using a telephone, people are interested in talking to people, not in dialling numbers. In modern office systems, the system will keep the list of telephone numbers and will also do the dialling. The user only has to specify to whom he wants to talk and let the system do the rest. According to the principle of multimedia communication he can do this in various ways: type the name of the person, point to the name in the telephone list or just tell it to the system via the microphone. The user will not have to pick up the receiver during the dialling process. When the person called picks up the receiver, his voice is heard through a loudspeaker, and this is when the person calling picks up his receiver.

6. Conclusions and outlook

- Tomorrow's office will be multimedia: both multimedia documents and multimedia human-computer-communication
- Systems will be more flexible and adapt to specific user needs and habits
- The user will be freed from the burden of knowing locations, numbers, etc., he says what he wants – and lets the system take care of how to do the job
- Knowledge-based systems will extend the use of modern technology from supporting routine-tasks to supporting more complex and sophisticated tasks
- With the integration of features like handwriting in electronic systems, office systems of tomorrow will be able to cope much better with office tasks, procedures and habits – thus leading from partial to complete solutions
- Modern desks will on the outside look more and more like those of the pre-electronic age. Horizontal displays will make traditional terminals disappear. The modern office desk will look similar to the desk of the 19th century clerk. The electronics inside, however, provides help and services the latter could not even have dreamt of.

Acknowledgements

I would like to express my appreciation to the members of the research department of TRIUMPH–ADLER for their participation in the work decribed in this paper. I am particularly indebted to my colleagures A. Fauser and R. Lutze for their contributions and discussions.

Literature

EGLT 78 ENGEL, GRAPPUSZ, LOWENSTEIN, TRAUB, An Office Communication System, IBM Systems Journal 18 (1979), pp. 402–431

ELNU 80 ELLIS, C., NUTT, G., ACM Computing Surveys 12 (1980), pp. 27–60

FISC 82 FISCHER, G., Mensch–Maschine–Communikation (MMK): Theorien und Systeme, Habilitationsschrift Universitaet Stuttgart, 1983

MSIS 83 MOKAWA, SAKAMURA, ISHIKAVA, SHIMIZU, Multimedia Machine, IAP 83, pp. 71–77.

OTPE 83 OTWAY, H., PELTU, M. (eds) New Office Technology. Human and Organizational Aspects, Frances Pinter Publ., London

UFB 79 UHLIG, FARBER, BAIR, The Office of the Future, North-Holland

WOEHL 84 Automatic Classification of Office Documents by Coupling Relational Data Bases and PROLOG Expert Systems. Proc. 2nd Conference on VLDB, Singapore, 1984

A retargetable debugger for the Karlsruhe Ada system

Peter Dencker

Hans-Stephan Jansohn

Table of contents

1. Introduction

During the development phase of a program and – as well – during the maintenance phase it is very important that programming errors can be quickly

* Ada is a registered trademark of the U.S. Government (AJPO).

located and removed from the program. Therefore a programming environment should very well support the analysis and location of errors. An appropriate means is an interactive debugging tool. The purpose of the debugging system described in this paper is to support the analysis of programs translated by the Karlsruhe Ada Compiler.

However, the Karlsruhe Ada Compiler is easy to retarget and easy to rehost. Several versions are already available (VAX/VMS, SIEMENS/BS2000, and SIEMENS/BS2000–MC68000 Cross compiler), several others are planned (Nixdorf 8890/VM/ESX, Perkin Elmer 3200/OS32, and VAX/VMS–LR1432 Cross compiler). Therefore it is very important that the debugging system is equally well retargetable and rehostable.

The scope of a certain version of the debugger are all programs (including tasking and real-time applications) translated by the corresponding version of the Karlsruhe Ada Compiler. This implies that for each version of the Karlsruhe Ada Compiler a corresponding version of the debugger is available.

The compiled Ada programs under test may either run in the environment of the host system, which contains the compiler, the program library, and the debugger or they may run on some target system (including embedded system) with more or less restricted communication channels to the host system.

The debugging system will be developed in two major stages:

(1) post mortem dump analyser

(2) online debugger

The first stage, the post mortem dump analyser, is partially implemented. The other will be implemented within the next year.

This paper is organised as follows. The concept of post mortem dump analysing an online debugging are explained in the next two Sections. In Chapter 2 the general requirements and capabilities of the debugger are listed to set up the debuggers functional range. In Chapter 3 and 4 we show a debugger design which suits the requirements and capabilities established in Chapter 2.

1.1 Post mortem dump analyser

The basic idea of a post mortem dump analyser is that when the program stops with an error, in Ada e.g. when the program is abandoned because of an unhandled exception, the contents of memory is saved on file, the dump file. Then the post morten dump analyser examines this dump file and extracts information valuable for the programmer in order to locate the programming error. In this way, for instance, the values of variables and parameters can be inquired. To this end the post mortem dump analyser needs information about the runtime organisation and information from the compiler and the linker/loader which is stored in the program library.

1.2 Online debugger

An online debugger has all capabilities of a post mortem dump analyser. The difference is that the debugger communicates online with the program under test. If the program under test is interrupted control is transferred to the debugger. It then may examine the contents of the memory of the interrupted program in the same way as the post mortem analyser examines the dump file. Afterwards execution of the program can be resumed.

Additionally the online debugger allows the programmer to change the state of the interrupted program (e.g. change the values of variables) and to control the execution of the program (e.g. by means of breakpoints).

2. Requirements on the debugging system

Besides the global requirement of retargetability we impose several other general requirements on our debugging system.

(1) No recompilation must be necessary in order to run a program under control of the debugging system. This is important for transient errors in real-time programs.

(2) The ability to debug a program shall not impose overhead on the generated code. This means that the program must not be specially instrumented for debugging. This allows full debugging support even in a production environment.

(3) The executable code and data shall lie at the same memory locations regardless of whether the program is debugged or not. This is necessary to avoid errors which are due to placing the code at certain locations in memory.

(4) The debugger must provide a source level user interface. This means that all user interactions with the debugger are in terms of the Ada program. In this way, the programmer should have the impression that his program is run on an Ada machine.

(5) Access to low level information shall also be possible because Ada allows to interface the non Ada world through the pragma INTERFACE, through machine code insertions, and other low level features. Nevertheless for most users and in most situations this information is not needed to understand the state and execution of an Ada program.

(6) The debugger must be applicable in a host-target environment. It may e.g. run remote on the host system as a post mortem dump analyser, or as an online debugger if an online communciation channel exists between host and target, or directly on the target if the program library and sufficient resources are available on the target.

These general requirements are supplemented by the capabilities to analyse and interact with the program under test given in the next two Sections.

2.1 Analyser capabilities

(1) Inspection of the source code.

(2) Inspection of the values of objects (variables, constants, parameters). The values are displayed in a form consistent with their representation in the source. For instance a record is displayed as a named aggregate. Access objects get a special format.

(3) Inspection of the status of currently existing tasks.

- name and type of task
- state (executing, suspended, completed, . . .)
- point of execution (in the source)
- tasks with outstanding calls on all entries of each task
- set of all open alternatives of select statements per task.

(4) Inspection of the stack of subprogram activations for each task.

(5) Navigation through the dynamic program structure (e.g. walk back, task inspection). In parallel the visible environment is changed.

2.2 Interaction capabilities

(1) Placement of breakpoints at Ada statement and declaration boundaries.

(2) Assignment of actions to breakpoints which shall be taken upon reaching a breakpoint.

(3) Modification of data.

(4) Interruptability of the program in execution by the user in order to inspect it at arbitrary moments.

(5) Stepping through the program in granularity of declarations, statements, calls, or task interactions.

(6) Tracing of executed declarations, statements, calls, or raised exceptions.

(7) Assignment of breakpoints to certain (or any) exceptions.

(8) Assignment of breakpoints to blocks under the condition that they are left by an exception.

(9) Hold (all) task(s), release task.

In the following Chapters we show the design of a debugger for the Karlsruhe Ada system that satisfies these requirements.

3. Inspecting the program state

Within this Chapter we assume that a program has been interrupted in some program state. We show how the debugging system can interpret this program state in order to answer the inquiries of the programmer.

Since we do not want to impose overhead on the execution of the program we do not collect information during its execution. Hence, the current program state is the only information the debugging system can access concerning the execution of the program.

The debugging system can, however, consider further information

(1) gathered by the compiler when compiling the individual compilation units of the program in the program library,

(2) from linking the program and

(3) about the runtime organisation.

The runtime organisation may very strongly depend on the target machine. However, these implementation details can be hidden behind a machine independent interface as long as we consider the same compiler.

The debugging system must access this information and derive from it a lot of data valuable for the programmer during the analysis of a particular error situation. The debugging system should work interactively to allow the inquiries of the programmer to depend on the results of earlier ones.

In the following we discuss in detail how the debugging system can answer several characteristic questions concerning

(1) the currently executed source statement,

(2) the calling hierarchy and

(3) the values of objects.

3.1 Retrieving source statement information

During debugging there is often a need to determine the position in the source that corresponds to a given code address or vice versa. For instance, the programmer may want to know which source statement raised an unexpected exception or he may want to interrupt the program when a certain position in the source is reached.

Since we consider a language with separate compilation, a position in the source is not uniquely defined by a source line number. Instead, we need a pair where one component indicates the compilation unit of the position. On the other side, since single statements may be distributed over several lines and since several statements can be written on one line, source line numbers are not appropriate for identifying the positions of single statements.

In the Karlsruhe Ada system the individual compilation units are stored in their Diana representations in the program library. The Ada source can be retained from the Diana representation of a compilation unit. Hence, it is suitable to use references into the Diana representations for identifying positions in the source.

For the following it is not of interest how the positions in the source are identified. We assume only that they can be uniquely identified. We call such an identifier a source position.

A common technique for determining the source position of the currently executed statement is to use a variable (in the runtime system) which always holds it. It is updated when the current statement changes.

This method has three great disadvantages:

(1) The code size as well as the execution time of the program are increased significantly (25–50%). This fact causes most programmers to switch off the generation of source statement information. So debugging of the programs is not further possible without recompilation.

(2) This method allows only to determine the position of the current executed statement. By this way, the name of the statement which e.g. called the subprogram containing the current statement cannot be obtained without storing it in the invocation frame. Further it is not possible to determine the code address(es) corresponding to a given position in the source (e.g. for implanting breakpoints).

(3) For rather complex statements more detailed information is needed for selecting the erroneous part of the statement. This would, however, result in much more overhead.

We propose another method which does not possess these disadvantages: The compiler builds tables which contain the mapping of code addresses to source positions and stores them in the program library. The debugging system interprets these tables for determining the position in the source which corresponds to a given code address. Since the relevant addresses are (normally) available during the execution of the program no additional code must be generated. For instance, the program counter contains an address corresponding to the currently executed statement, the return address is already stored in the invocation frame for subprogram calls. Hence there is no runtime overhead and the code size as well as the execution time is not increased.

Now we discuss how the tables containing the mapping of the code addresses to statement names are constructed. Since we consider a language with separate compilation facilities the compiler does only know the mapping of module relative code addresses to source positions. Therefore for each compilation unit one table is built containing the module relative information.

When linking the program, the linker computes the absolute addresses of the code modules resulting from the individual compilation units. These addresses together with the module names are usually written on some file. We call it the linker listing.

Given a particular code address the debugging system looks into the linker listing and determines to which code module it belongs. In presence of code overlay additional information is required from the program state in order to resolve the resulting ambiguities. Then it computes the module relative code address and obtains the corresponding source position by inspecting the table built by the compiler for this module.

If a source position is given the code address(es) corresponding to this source position can obviously be computed in a similar manner.

We have implemented this method in the Karlsruhe Ada system. By a simple compactification method we could reduce the size of the compiler generated tables to about 20% of the size of the generated code. So the size of these tables does not cause any problems. It is even less than the increase in code size which would result from the other approach.

3.2 Determining the dynamic calling hierarchy

In order to analyse an error situation the programmer has to inspect the source. The debugging system can support him by selecting those parts of the source which are associated in some way with the error situation. In the following we call a position in the source together with the context of the current program state a location. Starting from a location the programmer can inspect the static environment directly in the source.

However, this does not suffice. Additionally the programmer is interested in how the program came to a certain location. Especially, he wants to know the subprogram or entry calls which led to this location.

For Ada programs we therefore distinguish the following locations indicating the dynamic context of the interrupted program:

(1) Each currently existing task or the main program is interrupted at some location. This location is called the current location of the task or the main program.

(2) A location may lie within a subprogram. The location where this subprogram was called, is referred to as the calling location of it.

(3) A location may lie inside an accept statement. The location of the corresponding entry call is the current location of the task issuing the entry call. Hence, it suffices to know this task. We call it the calling task of it.

In order to be able to inspect each existing task the debugging system must know them all. Since in Ada each task depends on some master, all existing tasks can be enumerated if for each master all tasks which depend on it are known.

This leads to the following definitions:

(1) The set of tasks which depend on library packages are called library tasks.

(2) The set of tasks which depend on the main program or on a task t or on a block or subprogram currently executed by the main program or the task t are called the dependent task of the main program or the task t.

The debugging system provides means for accessing these tasks, their current locations, the current location of the main program and the corresponding calling locations and calling tasks.

By this way, the programmer can get a complete overview of the current program state. Since the program structure can be very complex we introduce for convenience the notion of the actual location, i.e. the location actually inspected by the programmer. The debugging system provides operations for inspecting the actual location and for moving to another location thus making this location to the actual one. The static context of the actual location can be inspected directly. This applies to the inspection of objects discussed in the next Section as well.

For implementing these operations the debugging system needs detailed knowledge of the runtime organisation, especially of the tasking implementation. Therefore the implementation of these operations does strongly depend on the target machine. However, since we consider always the same compiler, these machine dependent parts can be hidden behind machine independent interfaces. By this way, the debugging system remains portable.

3.3 Accessing objects

The possibility to inspect values of objects means great support for analysing error situations. When, for instance, the program was abandoned with CONSTRAINT_ERROR because of a subrange violation the question concerning the relationship between a value and the subrange bounds arises. This example shows that accessing an object must also include the possibility to inquire certain attributes of the object or of its type, e.g. its constraints, if any.

The debugging system must solve three totally different problems when it allows the programmer to access the value of objects:

(1) The programmer will, of course, enter the name of an object in terms of the source language. The debugging system must then identify the object denoted by the name and find out where this object is currently stored. We refer to this place as the address of the object.

(2) The value of an object is represented by some bit string stored at its address. The debugging system must be able to interpret this bit string as a value of the type of the object.

(3) If the object is constrained, the debugging system must be able to access the values of these constraints. For instance, for any array, the lower and upper bounds of all dimensions must be accessible. This also applies to other attributes of the object or its type.

These three items are discussed in detail in the subsequent paragraphs.

3.3.1 Locating objects

In order to ease the navigation through the current program state we introduced in Section 3.2 the notion of the actual location. We define here in terms of the source language which objects are accessible to the programmer in the actual location. So, by moving through the program (as described in Section 3.2) all objects existing in the program can be made accessible.

An object or package which is visible at a location or which could be made visible there by qualification is accessible at this location. Additionally, all library packages and all objects or packages which are declared in packages accessible at this location and their bodies are accessible at this location as well.

By this way the programmer may access (beside the objects visible at the actual location) objects which are hidden by an inner declaration and objects declared within the bodies of visible packages.

It seems necessary to allow the programmer to inspect the implementation details of packages during debugging although this violates the information hiding principle of the language. Perhaps it would be more appropriate to establish a protection mechanism which allows only the implementor or maintainer of a package itself to inspect its body. This could be achieved e.g. by the debugging system asking the programmer to enter the appropriate password when he wants to access hidden information or by hiding the information from the program library in general.

The locating of objects works in several steps:

(1) The name entered by the programmer is analysed whether it denotes an accessible object. Ambiguities in naming caused by the source language, e.g. by overloading or hiding, must be resolved during this step.
(2) The definition of the object in the Diana representation of the program is searched.
(3) The program library is accessed in order to obtain addressing information for this object. This results in a pair consisting of the name of a frame and the offset of the object within this frame.
(4) To compute the address of the frame further information must be retrieved:

   - If the frame is allocated statically, its address can be obtained from the linker listing.

- If the frame corresponds to the invocation frame of a subprogram or a block, its address can be obtained from memory of the interrupted program: the display vector or the static links must be inspected.
- Otherwise, the frame address is stored in a pointer object. The address of this object is computed in the same manner.

(5) Finally, the address of the frame and the offset of the object are added.

So, the address of the object is obtained.

3.3.2 Representation of objects

After an object has been located the debugging system must be able to interpret the bit string stored at its address as a value of the type of the object. For this purpose the debugging system must know the representation the compiler has chosen for this object.

Since the representation of objects is fixed by the type mapping module of each compiler individually fitting to its own runtime organisation, we have here again the situation that the debugging system depends strongly on the compiler, but not so strongly on the particular target machine.

Some characteristic data about the representation of objects are stored in Diana, but this information does not completely describe the representation. For instance, for integer objects the debugging system must additionally know whether the numbers are stored as binary numbers or as binary coded decimals, whether a sign bit is present, whether negative numbers are stored as one's or two's complement, etc.

This shows again that the debugging system must access the program library to get complete information. Additionally some target machine dependent issues must be considered which may be hidden behind a machine independent interface.

3.3.3 Constraint information

Constraints may but need not be static in Ada. Hence, for some constraints additional objects must be introduced which hold their values. The information which object holds which constraint is stored in the program library and must be inspected by the debugging system. Once the objects holding the constraints are known, Paragraphs 3.3.1 and 3.3.2 apply.

For the values of other attributes of the object or its type similar remarks apply.

4. Debugger interactions

In this Chapter we discuss the implications of the interaction capabilities for the debugger-compiler interface.

The debugger is conceptually a (monitor) task with its own stack. If it runs as a post-mortem dump analyser no interaction between debugger and program under test is necessary except for the generation of a dump (file) on the programs side and the identification of that dump on the debuggers side. If the debugger is on line with the program under test different possibilities for its implementation are seen.

(1) The debugger is loaded together with the program under test getting its own memory in the address space of the program under test.
(2) The debugger runs in a different address space (parallel task with message passing).

The latter possibility has been chosen because it allows to handle host-target debugging in the same manner as host-host debugging. In the former case the debugger runs as a task on the host and the program under test on the target. In the latter case both run as different operating system tasks on the host computer.

For the location of breakpoints corresponding to source positions it is necessary to have the compiler build tables which contain the mapping of source positions to code addresses. The code to source mapping proposed in Section 3.1 is not sufficient because it does not allow to retrieve from a source position the code address which is executed first. The tables containing the source to code mapping are constructed in a similar manner to those containing the code to source mapping.

In order to set breakpoints on exception or tasking events the debugging system must have detailed knowledge of the runtime organisation and tasking implementation.

For the modification of data the same information as for accessing objects is required. Additionally the debugging system must be able to interpret a string given by the user as a value of some type and transform it into a bit string to be stored at some objects address.

5. Conclusion and prospects

In this paper we described the design and partially the implementation of a retargetable debugging system for the Karlsruhe Ada compiler. The most essential implication is that there is no need to have two versions of the program: one for debugging and one for real usage. Otherwise it would be very difficult, perhaps impossible, to guarantee that the instrumented version will produce the same error situations. On the other hand, debugging code must not be incorporated into the program to avoid overhead for the normal execution.

Hence, the debugging system has to work for programs which are not specially instrumented for debugging. All knowledge about the execution of a program must be retrieved from its memory when interrupted. Additionally, the debugging system may access information collected by the compiler, the linker and information about the runtime organisation of the programs.

As has been discussed in Chapter 3 the target dependent features of the debugger can be concentrated in a few machine dependent packages making it easy to retarget the debugging system. The rehostability is given because the debugging system is written in Ada, as is the compiler.

We have shown that on this basis a reasonably working debugging system can be built: it provides all operations needed by the programmer in order to analyse an error situation.

A further advantage of this approach lies in its applicability to cross developed programs: The program is developed and tested on a host computer, but is used on some other machine, the target computer. If an error situation occurs during real usage, the program state (i.e. the contents of memory) can be written on tape and transported back to the host machine. There the environment is present for applying the debugging system (in the version for this particular target machine). By this way, error situations can be analysed even for real time applications where errors usually cannot be reproduced.

On the other hand this approach allows to 'field test' the program on the target with the debugger located on the host if an online communication channel exists between host and target.

5.1 Outlook

A shortcoming of most available debuggers (including the one presented) is that the program under test has to be stopped in order to analyse it. For some real-time applications this may be a serious problem. Therefore we are planning to develop debuggers which analyse the program behaviour in source level terms without disturbing or interrupting the running program on a target system. This is a most challenging effort with respect to real-time application programs.

Man–machine approach in ESPRIT projects

Thierry Van der Pyl

CEC/ESPRIT – ITTTF A25, 200 rue de la Loi B1049 Brussels

Design of MMI

Introduction

The design of MMI's must take into account all the contact points between the user and the system. It includes the issues dealt with by classical ergonomics and also 'non-classical' approaches. Several issues in this area have been considered as classical ergonomics, like:

- the Interface Language
  - ★ Commands
  - ★ Messages
  - ★ Menus
  - ★ Graphics
  - ★ . . .
- error and Help messages
- display on the screen
- authorised sequences of operations for the user
- on line documentation, teaching and possible recovery procedures in case of failures.

State of the art

A major shift of emphasis has taken place in the design of computer systems where the concept of usability has superseded the one of efficiency of machine usage already in a lot of application domains. This results in a greater interactivity of the systems with the user and a greater designer emphasis on dialogue (e.g. in CAD and CAI systems). Moreover attempts are made to separate out the dialogue aspects from the computational aspects and to make explicit in the system the representation of the user and the representation of the tasks to be achieved (see Fig. 41).

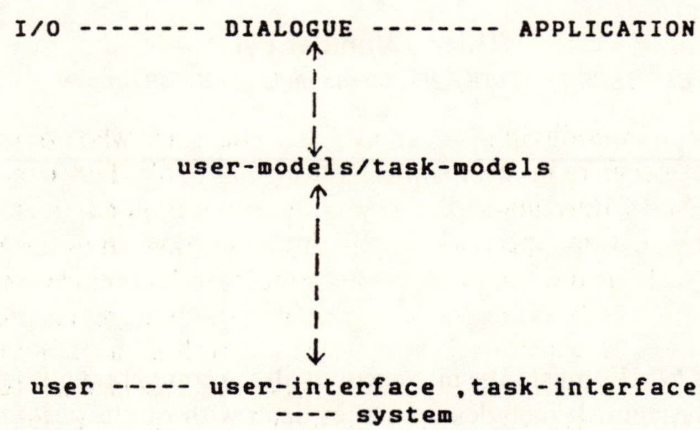

Fig. 41

Distinction was drawn between 'user-interface' and 'task-interface'. In the user-interface emphasis is given to the exchange of information between the user and the system. The user must get accurate and effective mental models of the system. This is done practically through the design of a command language, the selection of communciation devices and the generation of appropriate system messages. Errors are caused by mismatches between the user representation of the system and the designer model.

In the task-interface emphasis is given to 'what kind of support is provided by the system for the user task'. This implies development of reasoning strategies e.g. for diagnosis, of knowledge representation techniques e.g. to represent the knowledge of the actual user.

Evaluation of MMI

The evaluation of MMI is a very important and very difficult issue. Even objectives so well recognised as 'user friendliness', 'easy learning' are very difficult to define and to measure. Probably our knowledge is not sufficient to have systematic and proven methods to design interfaces and at the moment manageable and achievable objectives are more in terms of

> 'the system must respond in n seconds' or 'the training of the user must be less than m days'

than in terms of

> 'the system must be user friendly' or 'the system must take care of the user' . . .

Conclusion

It is likely that a multidisciplinary approach is necessary when one considers the number of issues to be tackled in the design of MMI. However the basic assumption that Information Processing is computational has led to a 'Cognitive Engineering Approach' of the design of MMI in order to provide the designers with the tools required to make their products more sensitive and responsive to the needs of the users. The results will be a greater productivity and a diminution of constraints on the user as well as a better quality of production. It might also result in a reduction of absenteeism and turnover of personnel. As a consequence the encapsulation of the concept of 'user-friendliness' in the application packages and in computer systems will increase their marketability.

MMI in the ESPRIT context

Generality

Esprit is a 10 years precompetitive Research and Development programme in Information Technology managed and funded by the Commission of the European Communities (CEC). The funding has been voted for 5 years (ESPRIT I) and amounts 750 MECU. The second phase of Esprit (ESPRIT II) is under discussion. Fig. 42 indicates the 5 areas covered by Esprit together with their respective percentage of funding.

MEL, ST, AIP, OS, CIM stands respectively for Microelectronics, Software Technology, Advanced Information Processing, Office System and Computer Integrated Manufacturing.

Topics related to MMI are mainly addressed in AIP and OS. Whereas MMI in OS is more concerned by human factors and classical ergonomics in office environment, MMI in AIP is related to various aspects of Speech Understanding, Natural Language Processing, Dialogue and Process Control. In the following focus is given on AIP.

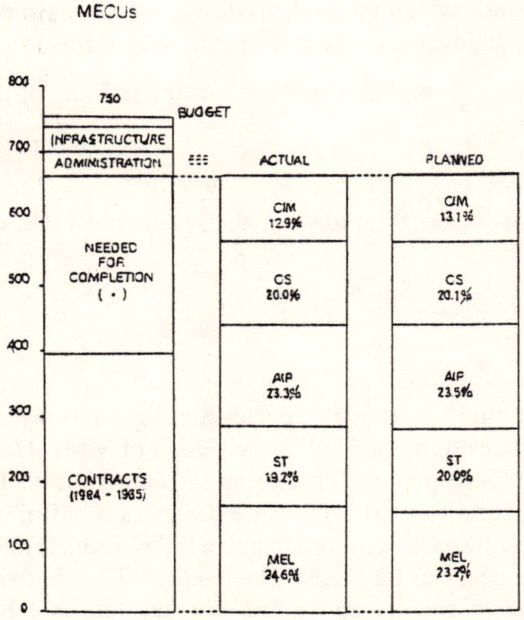

MECUs

800
750
BUDGET
INFRASTRUCTURE
700
ADMINISTRATION
ACTUAL
PLANNED
600
CIM 12.9%
CIM 13.1%
NEEDED FOR COMPLETION (•)
500
CS 20.0%
CS 20.1%
400
AIP 23.3%
AIP 23.5%
300
ST 19.2%
ST 20.0%
200
CONTRACTS (1984 - 1985)
100
MEL 24.6%
MEL 23.2%
0

Fig. 42

MMI in AIP

As AIP is at the front-end of current research in Artificial Intelligence and Signal Processing the projects tend to be very ambitious. However in relation to the definition of Esprit as a Research and Development programme, milestones and integration points are defined in the projects and prototypes in specific application domains are to be demonstrated. This forces the consortia to evaluate the available technologies and to make the appropriate choices.

Out of the 44 on going projects in AIP, 11 address issues related to MMI at least partly. This represents 59 MECU out of the 142 MECU funding of AIP i.e. 41%. Fig. 43 gives an overview of these projects.
Each project corresponds to a path in the graph.

Projects 107 (LOKI) and 530 (EPSILON) are concerned by 'intelligent' access to data and knowledge bases possibly in Natural Language (NL) for P107.

Projects 311, 316 (ESTEAM) and 641 (KIWI) are more concerned by handling large amount of knowledge with possible dialogue for financial advice giving (P316) or decsion support (P641).

Project 393 (ACORD) aims at the construction and interrogation of knowledge base through NL and graphical dialogue with the user in a business environment.

Project 280 (EUROHELP) aims at the establishment of methodology for providing intelligent Help facilities for users of information systems. The application domain is the Unix mail.

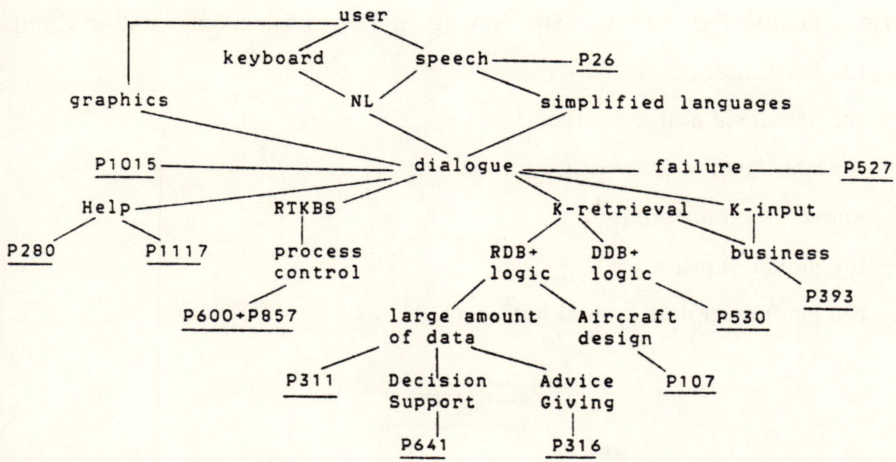

Fig. 43

Project 600 was concerned by Supervision and Control systems (S&C). A detailed study of the domain has been delivered highlighting the user requirements and the necessity of a cognitive approach. P857 aims at the implementation of the ideas developed in P600 for the design of intelligent graphics and knowledge base dialogue systems in S&C.

Project 26 (SIP) is mainly concerned by Speech processing and P1015 (PALABRE) by the interface speech/NL for interrogation of the yellow pages.

Annex

In this annex functional diagrams with short comments related to MMI part are given for some of the on going projects in AIP.

P280 (EUROHELP) Intelligent Help for Information System User (see Fig. 44).

For the design of the Help System the following aspects are considered:

- user modelling
- dialogue monitoring
- explanation
- teaching/coaching
- planning of task accomplishment
- domain knowledge
- use of advanced interface facilities.

Both the tutoring and explanation giving components need information about:

- the latest user action
- the actual state of the target domain
- the user level of knowledge and experience
- known misconceptions
- the global didactic goals
- the amount of help system initiative.

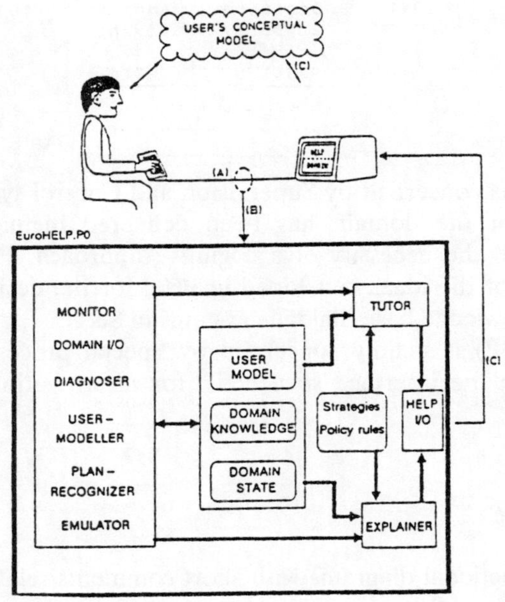

Fig. 44

P316 (ESTEAM) An Architecture for Problem Solving by Cooperating Data and Knowledge Bases (see Fig. 45).

The management of the dialogue has to do with the pragmatics of communication, and in particular with the understanding the goals which underly the messages produced by the user, and checking that the user appreciates the relevant implications of his own goals. A more general cooperation with the user is achieved by:

- helping the user to formulate his goals
- discovering problems the user is aware of and giving solutions
- convincing the user that the proposed solutions are adequate.

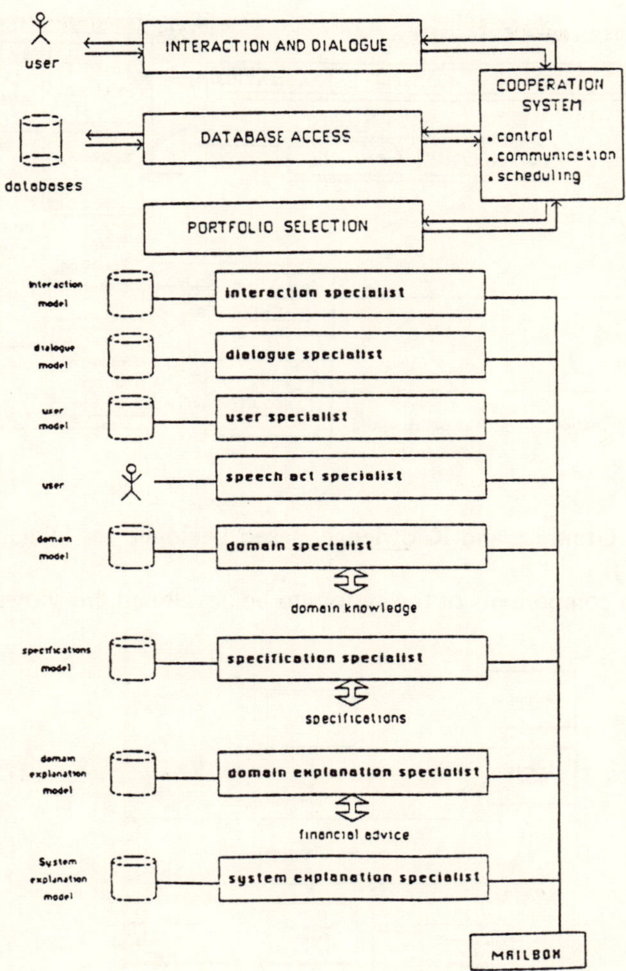

Fig. 45

P393 (ACORD) Construction and Interrogation of Knowledge Bases Using Natural Language Text and Graphics (see Fig. 46).

The knowledge base will be built from documents combining text and graphics and will be based on Prolog. The analysis of texts will be based on recent results in theoretical linguistics (discourse representation theory and functional grammars).

P527 Communication failure in dialogue: techniques for detection and repair.

In order to tackle the problem of communciation breakdown as broadly as possible a multidisciplinary approach has been adopted; it includes psychologists, computer scientists, linguists, and human factors experts.

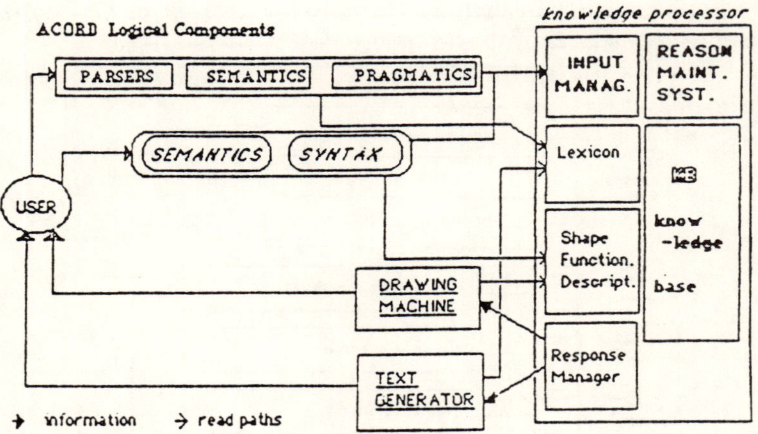

Fig. 46

P600/857 Graphics and Knowledge Based Dialogue for Dynamic Systems (see Fig. 47).

The main components of the system to be developed are shown in Fig. 47.

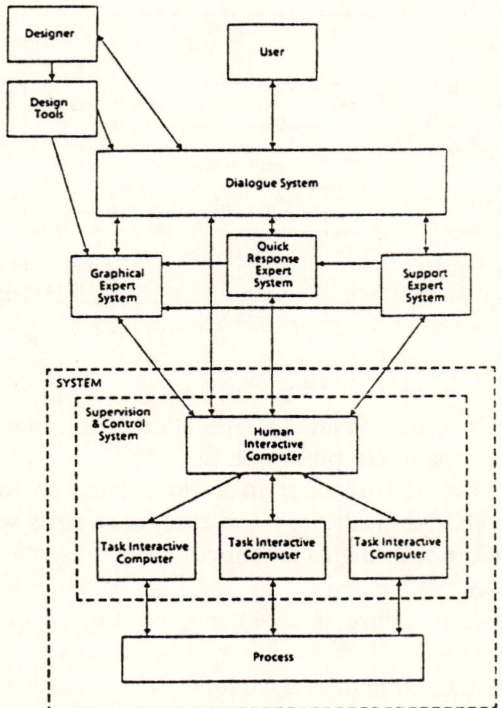

Fig. 47

The system surrounded by dotted lines represents an existing S&C system and an existing process. The rest is to be developed in the project.

P641 (KIWI) Knowledge Based User Friendly Interfaces for the Utilisation of Information Bases (see Fig. 48).

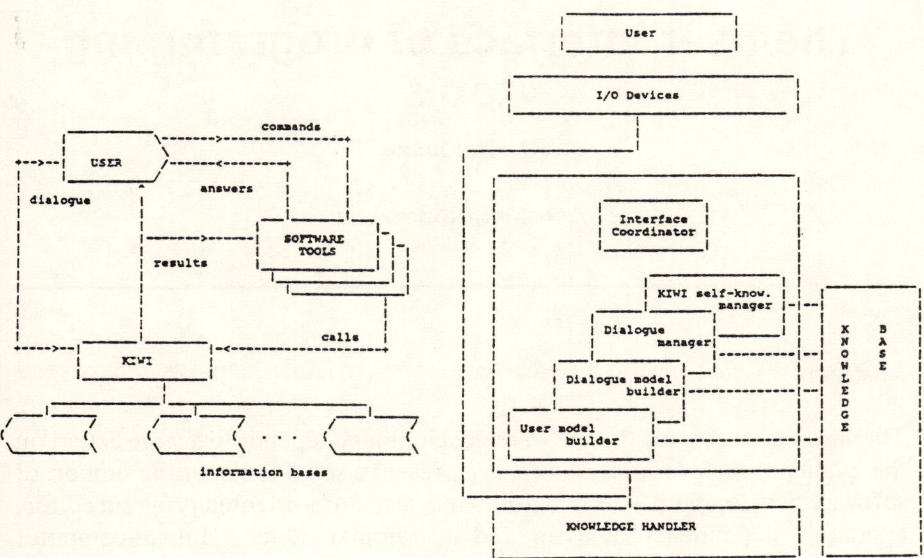

Fig. 48

P641 (KIWI) Knowledge Based User Friendly Interfaces for the Utilisation of Information Bases (see Fig. 48).

The User Interface (UI) assists users in the interaction with the KIWI system. In particular this software layer is expressed in terms of the knowledge representation formalism and is devoted to maintain a friendly and intelligent dialogue with users by allowing both their navigation through the KIWI knowledge base and an easy and assisted query facility on the information bases connected to KIWI.

Remarks: documentation on the on going projects can be found in the ESPRIT '85 STATUS REPORT published by North-Holland (1986) from which information of projects has been extracted.

The user interface of programming tools

J. Stelovsky

ETH Zürich, Switzerland

Abstract

The success of tools for the software development depends to a large extent on the quality of their user interfaces. We present a set of tools for the support of software development – a syntax-directed, structure-oriented program editor, a module dependency analyser and a grammar editor. Human-computer interaction aspects were the main focus during the design of these tools. In particular, we concentrated on the visualisation of the knowledge offered to the programmer.

1. Conventional software development: which tools are intelligent?

A conventional programming environment relies on few tools: a text editor, compiler linker and debugger. While new programming languages that support software-engineering aspects have emerged recently (Modula-2, Ada), the collection of support tools has not essentially changed for years [Macmeth]. Programs are developed as text, with an editor that has no knowledge about the programming language of the program itself. Since a text editor lacks the understanding of the program's structure and suggests the development of programs as sequence of characters, it also supports the wrong, bottom-up program design methodology: the programmer is concerned with the details before devising the overall structure. This disadvantage has an adverse effect particularly in education.

The only conventional tool that has the knowledge of the programming language – the compiler – is reluctant to reveal it: it only corrects syntactical mistakes without suggesting how they can be improved. Moreover, since it tries to generate a structure from unstructured text, it can detect subsequent errors where there are none or interpret them incorrectly. Thus the programmer lacks the support at the time when he needs it most – during the editing process.

The problem of uneven distribution of knowledge among the software development tools became even more serious when modular languages emerged. Since modular design implies that a program is dependent on library modules whose definition may not be available, the overview of large software becomes tedious. This problem is worsened if version control is rudimentary and the chore of keeping library modules consistent is left to the programmer.

The current improvement in user interfaces is largely due to improvements in hardware: high-resolution screen and pointing devices such as the mouse. The functionality of the programming tools, however, has not changed considerably: While numerous syntax-directed, structure-oriented tools were described in literature (e.g. Mentor [Donzeau-Gouge et al. 1975], Cornell Synthesizer [Titelbaum & Reps 1981], Pecan [Reiss 85]), their impact on the software-engineering practice is minimal. We attribute this lack of success primarily to the deficiencies in the user-interfaces. Indeed, tools that are too restrictive (i.e. force the programmer to obey the syntax at all times) or lack fundamental concepts (e.g. scope rules) are unlikely to be accepted by the community of software engineers. Also tools burdened with superfluous complexity and consequently require expensive hardware are unlikely to be successful.

2. The research projects at ETH related to software development

The research activities at the ETH in the area of software development tools are centred around the programming language Modula-2 [Wirth]. The main effort concentrates on the components of a conventional programming environment: text editor, Modula-2 compiler, debugger, operating system. Recently, such an integrated package called Macmeth was developed for the personal computer Macintosh.

In this paper, we will describe another effort: the design of a set of tools for Modula-2 environment focusing on:

(1) structured editing of programs based on the knowledge of the Modula-2 grammar – the program editor 'Macpeth'
(2) editing the language definition and general program layout – the 'Grammar Editor' and
(3) visualising the global structure of large software – the module dependency analyser 'MDA'.

Each of these projects is based on the principle that software development tools must possess knowledge about the objects they are manipulating: in programming it is the syntax and semantics of the programming language as well as of the structure and interdependencies of the software. We believe that the success of such tools is closely related to the quality of their user interface, i.e. how the knowledge is visualised and how it is offered to the programmer.

Moreover, we are convinced that such tools will be most useful in the education, where a correct design methodology needs to be supported. Here, the emphasis must be on structured problem solving and algorithmic design rather than on learning syntax and special features of a programming language. The same principle that was applied in the design of structured programming languages (for instance Pascal) should also guide the design of tools for software development:

To promote desirable methods, make them easy to be used.

3. Program editor Macpeth

In [Stelovsky et al. 1986], we postulated some requirements for program development tools. We have discerned between two main aspects:

(1) the static representation of the current state of the software

(2) the dynamic component of further development.

The static aspect concerns the representation of the software. It should reveal its structure, in particular its modularisation, and support the overview by displaying interrelations among its components. The dynamic aspect is particularly important, since it should provide the programmer with the knowledge of the programming language and of the software elements already available while he develops a new program. Typical text editors offer only a textual program display and supply no or very limited dynamic facilities (such as template substitution in [Donegan & Polunsky 85]). To overcome these deficiencies, we implemented the program 'Macpeth' that should supply both, program representations in various level of detail, and dynamic guidance by offering the syntactical constructs and software components already defined.

3.1 Program representations

Macpeth is a syntax-driven, structure-oriented program editor. With Macpeth, programs can be visualized in three different ways: as text, as block structure and as a parse tree. They need not be typed in as text, but can be developed as a structure by choosing from a selection of symbols defined by the programming language or delcared identifiers.

Fig. 49 shows the screen layout during the development of a simple program. The program's text representation is shown in the 'program window'. A small black triangle – the caret – shows the insertion position. The 'selection window' shows all possible objects which can be inserted here according to the Modula-2 grammar. The narrow window on the left of the program window is the 'program scroll-bar'. It allows the user to scroll a large program to find the part he is interested in. The other narrow window – on the right of the selection window – is the 'selection scroll-bar'; it is used to scroll the selection window. At the bottom of the screen, the 'dialog window' echoes the command being executed and reports errors, warnings and other messages.

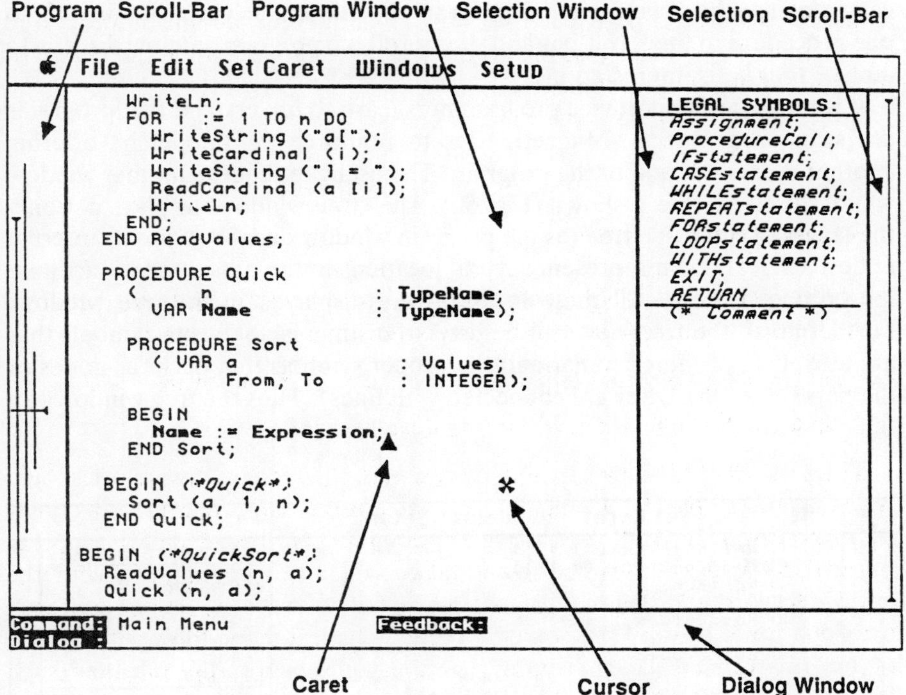

Fig. 49 *Screen layout in Macpeth*

The programs developed with Macpeth are parse trees. The textual representation in the program window is generated using display information stored in a grammar. Reserved words, punctuation signs and formatting instructions (i.e. indentation, tabulators, spaces and line ends) are attributes of this grammar (see also Section 4). As a consequence, a uniform layout of the program text can be generated automatically. In particular, indentation can represent the structure of the edited program. During the development of a program, incomplete and incorrect program parts cannot be avoided. Macpeth displays such places with bold font (Fig. 49) and provides a facility to search them.

The scroll-bar is an established concept in window-oriented user interfaces. Usually, however, it either contains no information or shows only the relative position of the window's contents in the document. Macpeth uses this area to display the block structure of the current program. The entire length of the scroll-bar represents the whole program and the leftmost line (with triangles at the ends) shows which portion of it is shown in the program window. The other vertical lines symbolise procedures (and modules) and visualise their nesting. The short horizontal line shows the current location of the caret. This 'block representation' is particularly useful for the experienced programmer, who can recognise the structure of his program. Moreover, he can point directly to one procedure in the scroll-bar and thus scroll the program window directly to a place that he is interested in.

It is important that novice programmers learn to understand that programs are structured objects. Macpeth tries to reinforce this model by offering another representation of the program. The user can request another window where the parse tree is shown (Fig. 50). This 'tree window' displays not only the leaves of the parse tree (as the program window does) but also its internal nodes. The caret can represent several locations in the parse tree to facilitate the editing process. All these locations are displayed in the tree window. Furthermore, the tree view can be used to distinguish between symbols that are attributes of the grammar and the proper symbols that occur as nodes of the parse tree (the latter are connected with lines). Thus the tree window can

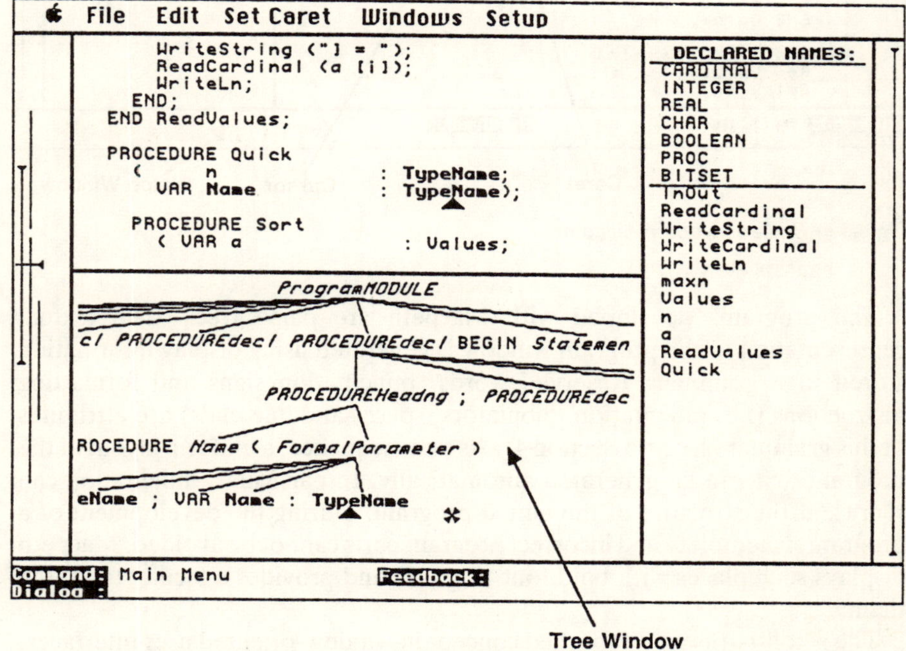

Fig. 50 *Parse-tree representation of a sample program*

help the user to understand in the structure of the program, the grammar of the programming language, as well as the behaviour of the editor.

Since the tree window always shows the environment of the caret, all three program representations have consistent contents. Since the block and tree representations are not always necessary, they can be hidden upon request to simplify the screen layout.

3.2. Editing programs

In the selection window, Macpeth provides the knowledge of the programming language and of the objects already defined. It displays the menu of symbols that can be inserted at the current location of the caret here. In Fig. 49, to give an example, the caret is located behind an assignment statement and the selection window offers all statement kinds. The user can point to one of them and click the mouse button to insert it. Fig. 51 shows the screen after the 'For-Statement' symbol has been selected. Notice that several additional symbols were (e.g. Expressions) inserted as prescribed by the Modula-2 grammar. Moreover, the program window displays them together with the necessary reserved words and punctuation signs. As a consequence, the programmer is freed from substantial typing effort. Furthermore, ominous 'missing semicolon' errors are eliminated.

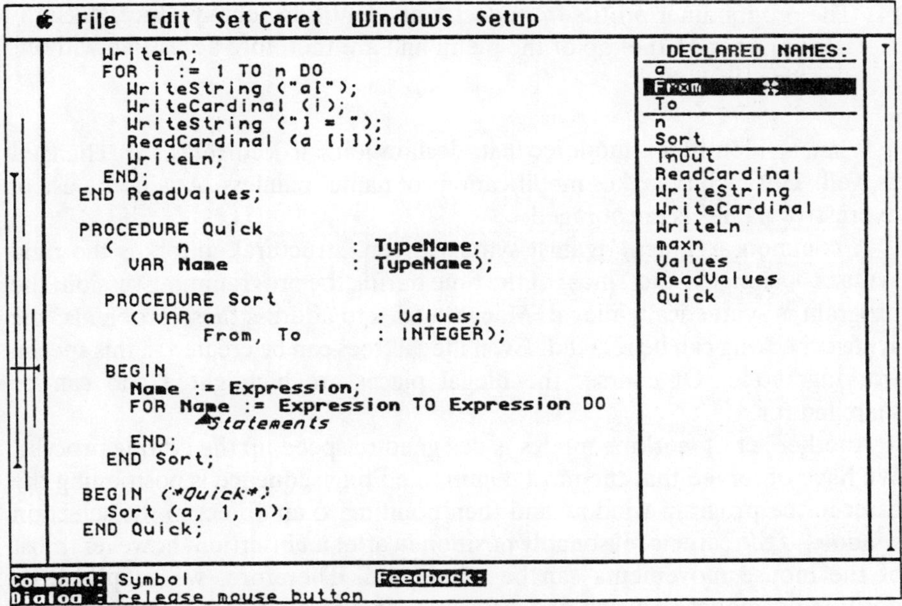

Fig. 51 *The sample program after a for-statement was inserted*

This demonstrates an important principle in the design of user interfaces for tools with strong consistency checking:

Offer correct choices instead of letting the user make mistakes and correct them later on.

This way the consistency constraints will be perceived not as a nuisance, but rather as a help. Suggesting syntactically legal constructs is particularly important for beginners: They need not memorise the entire grammar of the programming language or consult the syntax diagrams.

In Macpeth, the support of the dynamic aspect of the editing process is not restricted to the syntactic help. When the caret is located on an identifier (as in Fig. 51), all previously declared identifiers are offered in the selection window and can be inserted with the mouse. Only the identifiers visible according to the scope rules are presented; the identifiers declared in different scopes are separated by horizontal lines. While the user profits from the fact that no typing effort is necessary and orthographical errors are eliminated, selecting identifiers from a menu has desirable methodological consequences:

(1) The representation in the selection window fosters the understanding of scope rules.

(2) The programmer is encouraged to use long, expressive names which make the program self-documenting.

(3) Declaring objects before they are used increases the efficiency of editing.

(4) The programmer profits from declaring identifiers locally. Local declarations appear on the top of the menu and are therefore accessible without scrolling.

When an identifier is modified in its declaration, all occurrences are changed as well. This facility makes modifications of names painless. Again, the use of expressive names is encouraged.

A common argument against syntax-driven, structural editors is the rigid syntax checking. In fact, most of the time during the programming session, the program is syntactically illegal. Macpeth tries to address these problems: the syntax checking can be relaxed. Even illegal trees can be created in this special working mode. Of course, the illegal places are highlighted and can be searched for.

Another set of working modes is designed to speed up the editing process. We have observed that the most common editing sequence is positioning the caret in the program window and then pointing to an object in the selection window. If the caret is reasonably positioned after an insertion, however, most of the mouse movements can be eliminated. Therefore, we supply three positioning modes. During the first one, the caret is left behind the last inserted symbol. This mode is useful when the global skeleton of procedures in a module or of statements in a statement sequence is being developed and the details are left unspecified. This mode corresponds to the stepwise refinement. In the second mode, the caret is placed at the first relevant place in the inserted structure. To give an example, Fig. 51 shows the effect of this mode after the 'for-statement' was inserted. This mode is the most efficient for small modifications and for specifying expressions. In the third mode, the illegal or incomplete places are automatically searched for. It is useful in the last stage of the editing session, when the programmer wants to remove all errors.

Typing 'a:=a+1' requires less effort than developing the corresponding derivation tree. This complaint concerning the lack of efficiency in structural editors is often made by experienced programmers. Macpeth supplies a parsing facility to alleviate this problem. The programmer can always type in text. If the caret is not an identifier or another node where text is expected, the characters typed in are inserted into a new comment. A 'Parse' command will then convert this text into a parse tree. In this way, Macpeth combines the advantages of textual and structural editors.

4. The grammar editor

Macpeth is a syntax-driven editor: the knowledge of the underlying programming language is contained in a grammar. This grammar is stored in a separate data file and read in at the beginning of the editing session. We have developed another tool that creates and modifies such grammars. The Grammar Editor enables the designer of the programming language to define the productions of the grammar. For instance, the reserved words, punctuation signs and the possible sequences of the terminal and nonterminal symbols can be specified with this tool.

Since the display attributes (such as indentation, blanks, line ends) are also defined in the grammar, the programmer can use the Grammar Editor to specify his preferred program layout.

To facilitate the editing of grammars, the Grammar Editor presents the user with a graphical, familiar user interface: the syntax diagrams. Fig. 52 shows the screen layout. The large 'diagram' window contains a portion of the syntax diagram for the production that defines how the symbol 'Compilation Unit' can be expanded. Since such a diagram can be rather complex, we have introduced a hierarchy within a syntax diagram. This hierarchy is based on another familiar concept: the bracketing symbols used in EBNF (Extended Backus-Naur Form). But instead of the textual, barely readable EBNF specification of the grammar, we use icons for the repetition (i.e. '$\{s_1|s_2|\ldots|s_n\}$'), selection (i.e. '$(s_1|s_2|\ldots|s_n)$'), option (i.e. '$[s_1|s_2|\ldots|s_n]$') and sequence (i.e. '$(s_1s_2\ldots s_n)$') constructs. Notice that the icons symbolise the kind of syntax diagram generated by the corresponding construct.

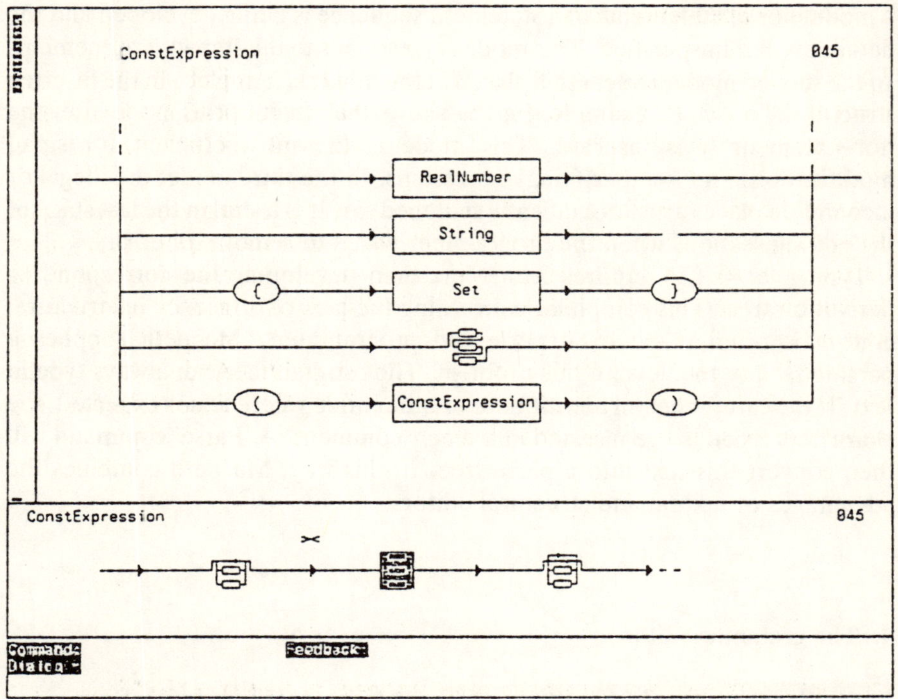

Fig. 52 *Iconic representation of a production in Grammar Editor*

The user can browse through the syntax diagram in two ways: Clicking on any of the icons shows its contents, i.e. goes one step deeper into the grammar tree. The smaller 'environment' window shows where the contents of the diagram window is embedded inside the production. Thus clicking in the environment window leads one level up and shows the environment in the diagram window.

Various editing facilities are provided in the diagram window: new branches can be inserted, existing ones (with the subtrees they represent) deleted or saved in the buffer. Within one branch, the display attributes, reserved words and punctuation signs can be modified. Modifying grammar can result in an inconsistent stage (e.g. when an undefined symbol name was used). Thus, we provide a search facility that finds inconsistent places.

Like Macpeth, the Grammar Editor knows about the objects which it manipulates: the productions of a grammar. It visualises them graphically and controls their consistency.

5. Module dependence analyser

The Module dependence Analyser MDA is another project under development at the ETH. Its goal is to visualise the global interdependencies in a large software at a higher abstraction level than Macpeth does: on the level of modules.

After the user specifies a collection of modules, MDA analyses their import lists and automatically augments the collection with the imported modules. Furthermore, MDA constructs a dependence graph ordered according to the dependence hierarchy (topological sort) and displays the module boxes on the screen (Fig. 53). The layout of the graph can be then modified (i.e. boxes relocated) preferably according to the logical structure. Additional modules can be included in the collection or excluded from it. The collection and its layout can be saved and reloaded later.

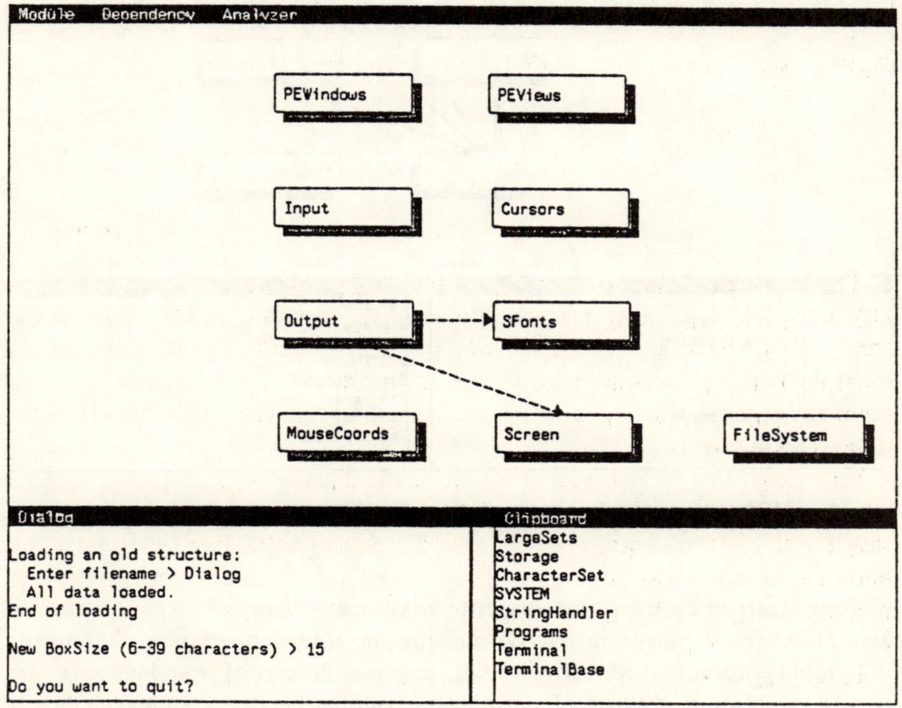

Fig. 53 *MDA shows which modules import 'Output'*

The main task of MDA is a selective display of the module dependences. The user can ask such questions as: (1) who imports a given module? (Fig. 53 shows the modules that import the module 'Output') or (2) which modules are imported by a given module? (Fig. 54 shows the imports of 'Output'). Even more important are queries concerned with indirect dependences. To give an example, the module M imports only M1 but the definition module of M1 imports other modules M2, M3, etc. Then M is also dependent on the modules M2 and M3. Such indirect dependences are important, since when a definition module is recompiled all directly and indirectly dependent modules must be recompiled. Also this task of recompilation is simplified: MDA can generate module lists and macro command files that recompile a portion of the collection in the correct order.

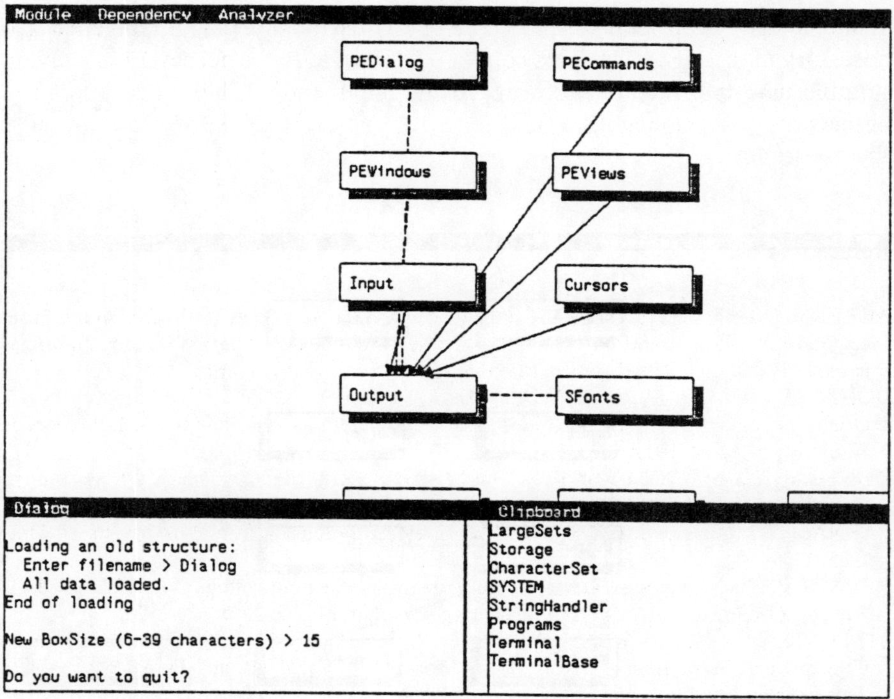

Fig. 54 *MDA shows modules that are imported in 'Output'*

Thus MDA helps the programmer to analyse and organise large software. It also enhances the overview by visualising the interdependences among the individual modules. Moreover, the tedious chore of making software consistent after a change of definition modules becomes an easy, mostly automatic task.

6. Implementation

Macpeth, Grammar Editor and MDA were implemented on the personal computer Lilith and then ported to the personal computer Macintosh. The software was written entirely in Modula-2. The portability was simplified due to structuring the software of all presented tools into three layers. The operating system layer was ported first, thus having quickly a running version. Then, the dialog layer was reimplemented to adapt the tools to the different user interface. The application layer, however, did not need to be changed at all.

7. Conclusion

While it can be argued whether the methods of artificial intelligence are necessary in managing complex software [Parnas 85]; it is necessary to provide suitable user interface to software development tools. They must know the objects they are managing, visualise them, and use their knowledge to guide the programmer.

References

ACKERMANN, D., STELOVSKY, J.: The Role of Mental Models in Programming: From Experiments to Requirements for an Interactive System. Interdisciplinary Workshop 'Informatics and Psychology': Visual Aids in Programming Passau/Schärding, May 1986.

DONZEAU-GOUGE, V., HUET, G., KAHN, G., LANG, B., LÉVY, J. J.: A Structure Oriented Program Editor: a First Step Towards Computer Assisted Programming. Rapport de Recherche No. 114, IRIA 1975.

DONEGAN, M. A., POLUNSKY, R. A.: Modula-2 Software Development System (User's Guide), Interface Technologies 1985.

PARNAS, D. L.: Software Aspects of Strategic Defense Systems. CACM Vol. 28, No. 12, Dec. 1985, 1336–1335.

REISS, S. P.: Pecan: Program Development Systems that Support Multiple Views. Proceedings 7-th Int. Conference on Software Engineering, Orlando, Florida 1985, 324–333.

STELOVSKY, J., ACKERMANN, D., CONTI, P.: Visualizing of Program Structures: Support Concepts and Realisation. Interdisciplinar Workshop 'Informatics and Psychology': Visual Aids in Programming Passau/Schäding, May 1986.

TITELBAUM, T., REPS, T.: The Cornell Program Synthesizer: A Syntax-Directed Programming Environment. CACM Vol. 24, No. 9, Sept 1981, 563–573.

WIRTH, N.: Programming in Modula-2, Springer Verlag, Berlin, 1983.

Management of software for large technical systems

H. Halling

Abstract

For the design, construction and operation of large technical systems, software plays an important role. In this presentation some aspects of managing software for such systems are discussed. The different classes of software during the project phases and within the different hierarchical levels of a control system are outlined and their relations to proper management are shown. In addition, the problems of purchasing software and estimating the required time, budget and manpower for a project are discussed. Emphasis is placed on practical aspects and examples are presented.

Introduction

During the last few years I have heard of many projects, where the top project management had developed an uncomfortable feeling about computer based activities and in particular about software. This was especially true for all projects where I was personally involved.

I think that this feeling has two main sources. Firstly, most of the project leaders never wrote software themselves, so that an understanding of the kind and amount of activities which lead to a useful product is missing. Secondly, it is a fact, that the control of the progress of software development, even on a high level with only a few milestones, is difficult. As a consequence, software and software producers cannot expect to find enough credibility and trust.

In my opinion, only the attempt to demonstrate that software products play an important role within a project and that software can be planned and kept under control in order to fulfil its tasks within the project, will lead to better understanding and trust. The following is an attempt to provide an overview concerning software and its management for large technical projects, keeping in mind that nowadays there is nearly no project activity where software is not involved.

Management of software includes planning, control and correction of all kinds of software activities through all project phases and levels of the control system. Activities are requirement analysis, evaluation of products, purchasing, organisation of training, hardware layout, installation and maintenance of packages, system tuning, preparation of user-friendly interfaces, structuring of databases, definition of internal standards and finally producing software products where necessary.

Clearly nobody can expect all these activities to be treated in detail in this presentation, but some of the interesting items are picked out, especially those which have changed recently, or which are in a phase of rapid development.

Project phases and their relation to software

Fig. 55 shows the important project phases and the software related to them. Software which is relevant for all phases is project management software, controlling deadlines and budget in correlation with the achievements of single activities and organisational support of the project.

The figure also shows that the software which is actually produced during the project and the tools and utilities needed for these activities are only a part

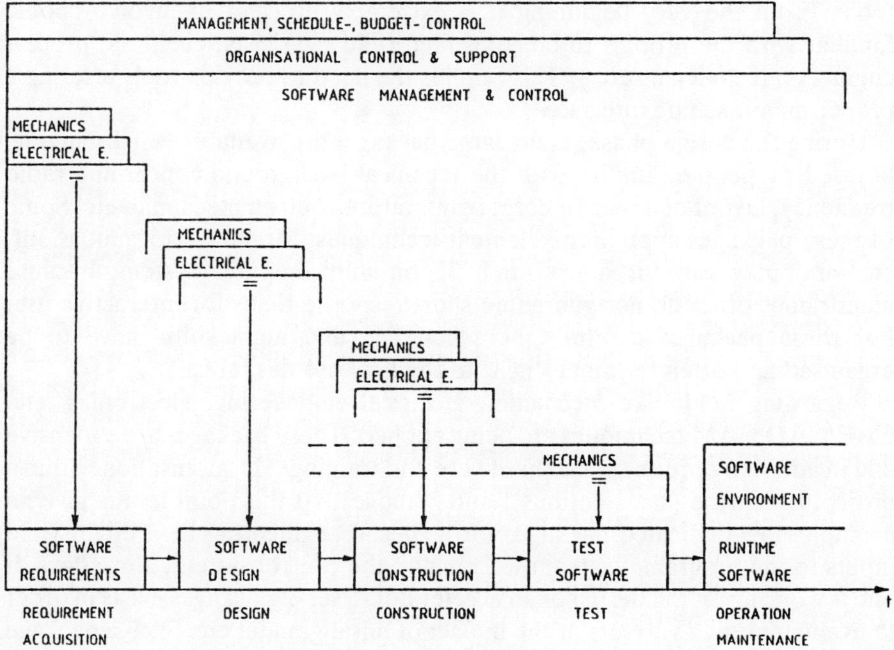

Fig. 55 *Project Phases and Software*

of all the packages involved. For other fields of engineering like mechanics, electrical engineering, thermodynamics, etc. packages exist for the different project phases. The results of such activities may be a valuable input for the process of producing software for the control system.

Some of the packages like the control of cost flow or deadline control must be accessible by everybody responsible for a part of the project during all of the project phases. This implies a common database and communication system. At this point, the question arises as to whether it is preferable to connect the relevant team members 1. via dedicated terminals to a dedicated management support system, or 2. to link the individual and most probably different data processing systems into a network running distributed management software and database systems. Solution 1 is simple and relatively cheap, but may later result in a variety of independent dedicated communication stars. Solution 2 may cause an incredible amount of work, may degrade the performance of smaller systems drastically and may end up with a different project where the programmers have to learn harsh lessons about standards and compatibility without achieving anything for the original project. Up to now such software is mainly running on mainframes, most of which support distributed terminal services. Another package used throughout the phases deals with organisational matters, including configuration control, documentation libraries, organisational communication etc.

Most of the software packages used in the earlier stages of a project are tools. From the very beginning it is extremely efficient to involve people familiar with the process to be controlled (like physicists, chemists, process engineers, technicians etc.). This implies the need to provide tools offering a proper man machine interface.

During the design phase, many large packages are available which can only be used by people familiar with the technical background concerning radio frequency, layout of a reactor core, temperature control, mechanics etc. Some of these packages apply finite element techniques, Monte Carlo simulations, etc. and may run for hours on a 32 bit minicomputer system, because mainframes often do not guarantee short response times for interactive use. For these packages careful time scheduling and night shifts have to be organised and often terminals 'near to the bed' are desirable.

Regarding fields like mechanics, electrical engineering, electronics, etc. CAE/CAD/CAM techniques are being applied. These packages are expensive and in addition require special hardware and training. Harmonisation within a project is delicate (see solutions 1 and 2 above). At this point let me give an example how dedicated packages for a special engineering field might yield inputs to the programmer team working on the control system. Nowadays, if one is working on the design of an accelerator, a set of packages exists in order to evaluate design variants or the impact of misalignment etc. Such a package can be used by the control people to calculate the accuracy of set points and the impact of deviations (Fig. 56). Thus the transfer of information between the

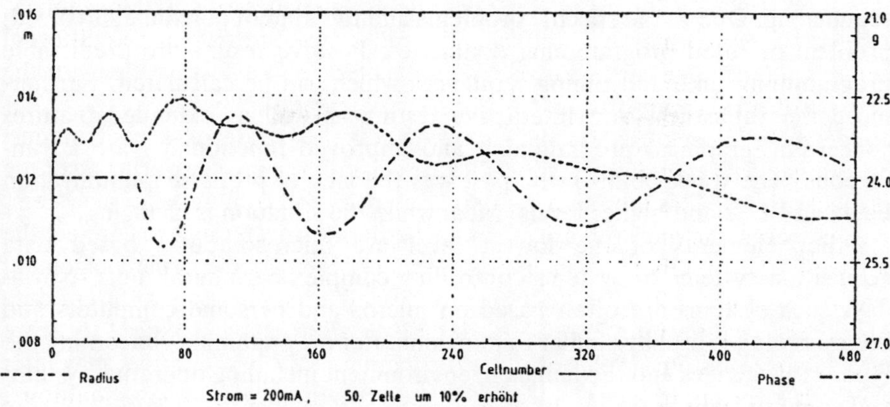

Fig. 56 *Simulation*

two parties took place by handing over a simulation package used by both to find out about different aspects. I am not going to discuss further details concerning the software production phases and their tools, because this is a familiar topic.

Structure of control systems and its relation to software

Due to a variety of reasons, the programmable elements at the various levels of a large control system are quite different (Fig. 57). On the lowest level programmable logic controllers (PLCs) are number one for interlocks and

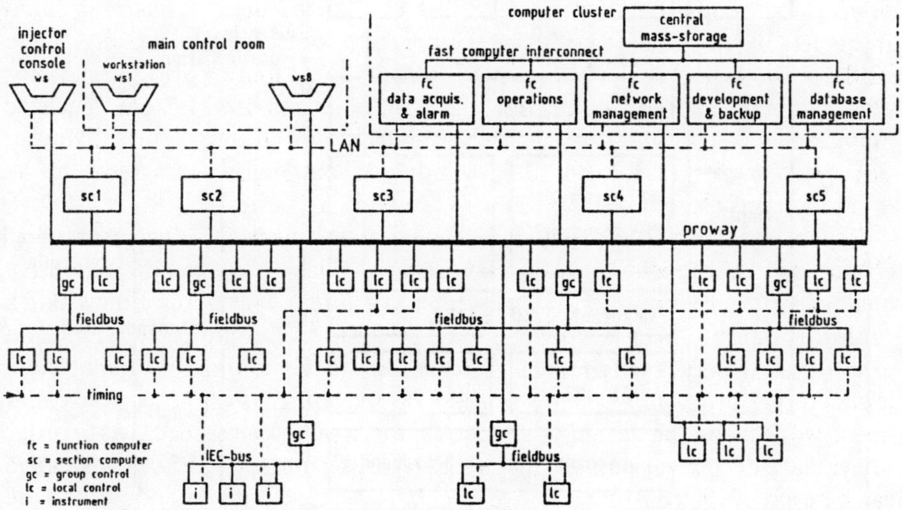

Fig. 57 *Structure of the Computer Control System*

sequencing. Due to the specific problems and the kind of personnel involved, problem oriented programming dominates. Positive results are predictable programming time and timing sequences which can be calculated. Furthermore the robustness and interactive testing as well as extended features concerning graphical programming and improved functionality are advantageous. The weak point in the past was the lack of proper communication between PLCs and higher levels. Meanwhile the situation is changing.

Other elements of the lowest level are microcomputer based data acquisition systems or systems controlling complex peripheral units. Nowadays such systems are often based on micros and personal computers and programming is usually similar to minicomputer programming concerning the types of languages and the language environment including operating systems and communications.

The connection of low level elements between each other or to group control units is done by serial field buses well suited to the harsh environment. As far as software is concerned, there is no standard communication at this level up to now.

The so called backbone communication system is on the way to be standardised at least to layer 4 (transport layer). This requires up to 100 kbytes of interface software but is the way to standard communication between low level elements and minocomputer systems which are the basis for the elements of higher levels (Fig. 58).

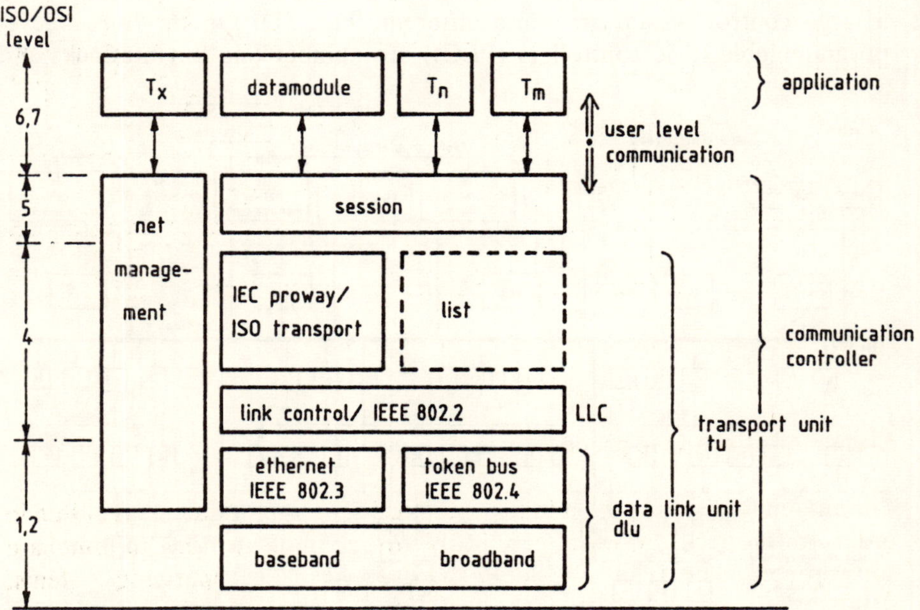

Fig. 58 *Standards for Communication*

Functionally the higher levels can be divided into three groups. Process control computer systems are responsible for higher level control and contain realtime software either stand-alone or with a proper operating system and realtime databases.

The second group is workstations or console systems mainly responsible for the high level man machine interface. This is the place for interactive graphics, menu libraries and in the future maybe even expert systems.

The third group is large minicomputer systems where the functional packages of the higher levels can be run. These include large application program packages, database management, all kinds of utilities and the communication software for intersystem communication and communication with higher level computers like mainframes or number crunchers. This minicomputer system runs the typical software for development, configuration control, longterm database management etc. (Fig. 59).

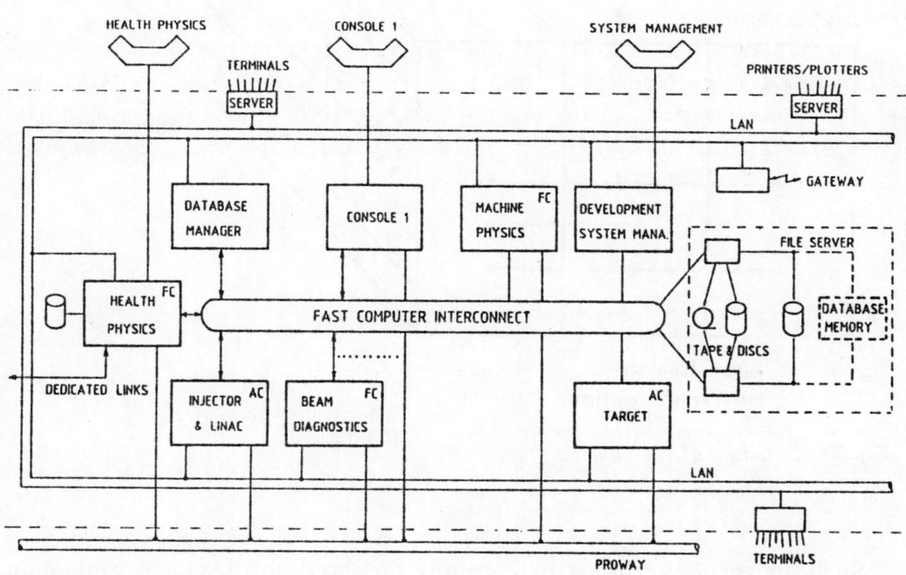

Fig. 59 *Central Computer Compound*

Evaluation of software products

The amount of manpower and the experience of the programmers required in order to produce software practically forces implementors to purchase software products. This is evident in case one needs operating systems, database systems, compilers, utilities or tools. But meanwhile even application software or at least parts of it can be purchase and tuned to meet the special requirements of a certain application. The only problem is that unlike hardware, a datasheet for software does not help too much. In general the products are so complex, that with present technology only a careful evaluation of the product leads to useful decisions.

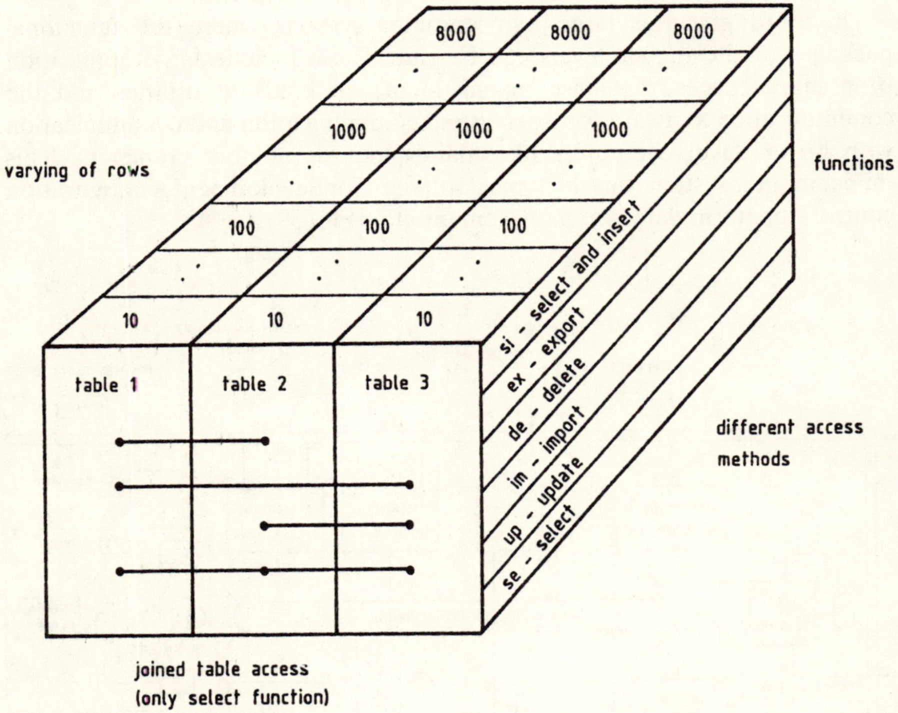

Fig. 60 *Structure of Testables*

Such an evaluation must be carefully prepared and result in evaluation specifications which contain what should be measured, how this should be done and in which order and how the results are to be documented etc. This ensures that different products are treated the same way. Figs. 60–64 show some examples concerning the evaluation of a relational database and a distributed operating system.

Tools are very difficult to evaluate. The ideal evaluation would be to run a small project applying the tools to be evaluated. The main advantage is that several programmers take part thereby reducing individual human preferences and also testing the cooperation between the group members. Definitely the introduction of Ada should follow this line.

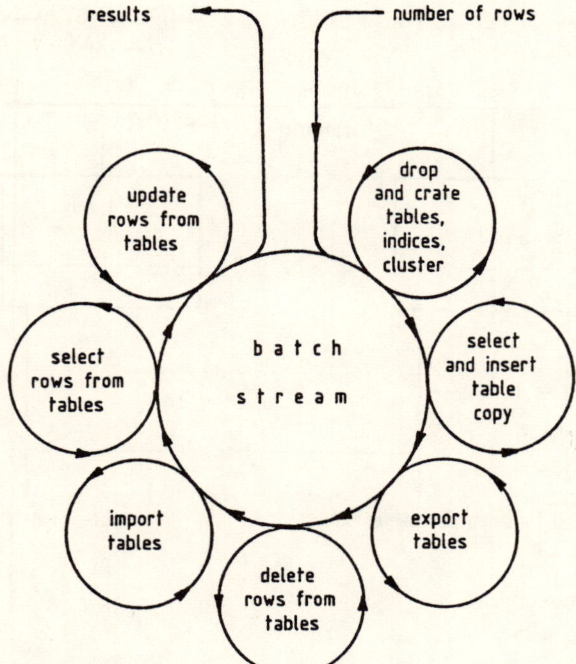

Fig. 61 *Test Sequences*

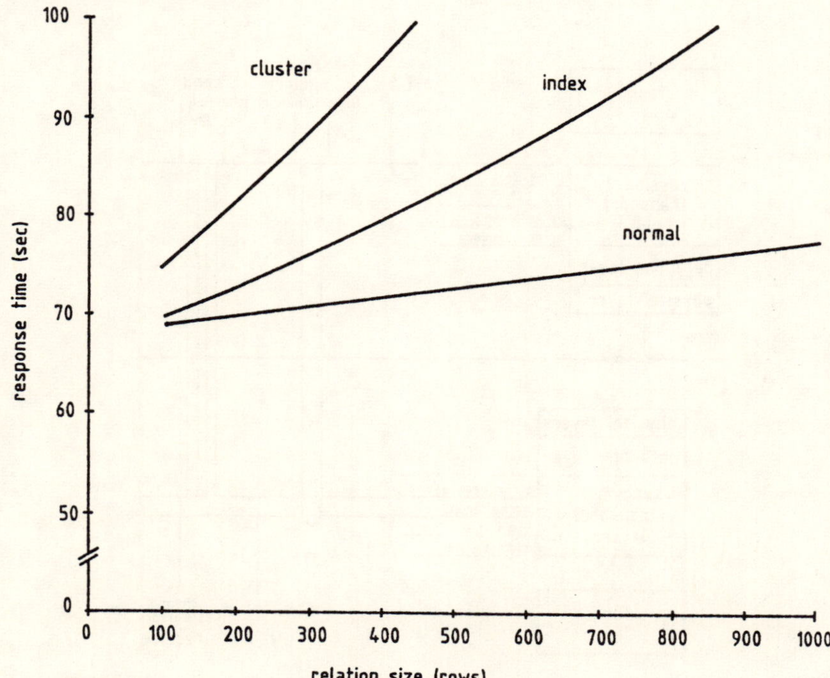

Fig. 62 *Results*

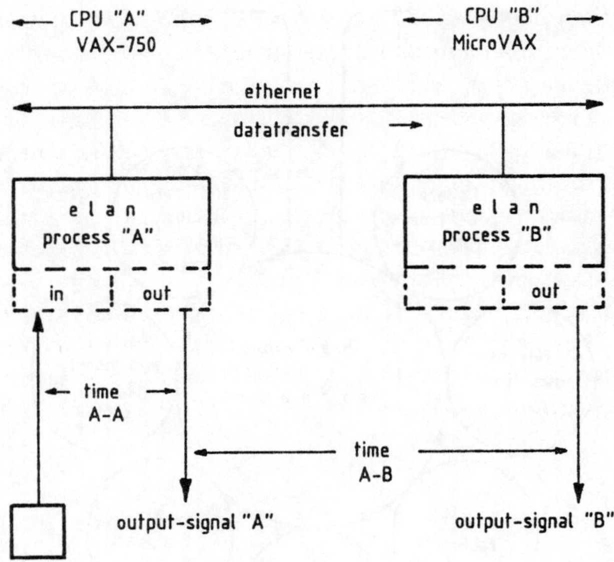

Fig. 63 *Test Setup*

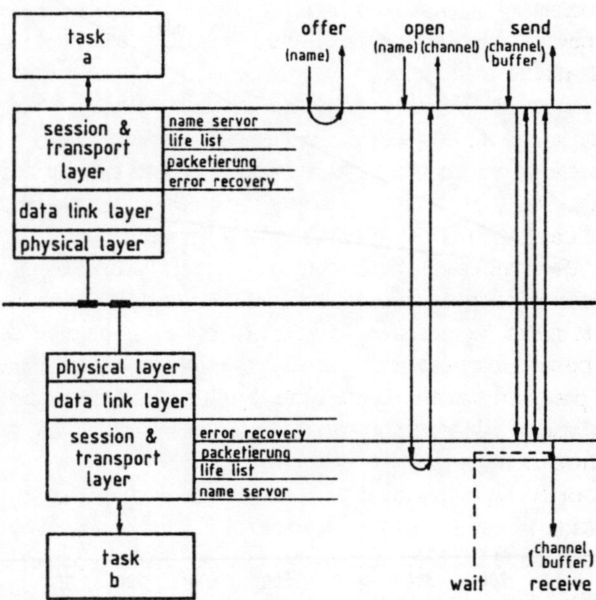

Fig. 64 *Test Description*

Programming languages: do not have the importance in practice as one might think when reading publications or listening to discussions at conferences. Actually most people working in big projects are familiar with several languages and except for very tricky programming, which is to be avoided anyway, they produce acceptable software. But this is mainly true for programming 'in the small'. For global design and construction, the language environment and the experience of programmers is much more important than the programming language. Meanwhile the first packages supporting programming 'in the large' are reaching the market.

As a matter of fact, years ago in the scientific community many application packages had been written in PASCAL. In the meantime most of the more successful have been rewritten in FORTRAN, in order to exploit better language environment and execution speed as well as portability.

Databases

There are different requirements concerning databases and their management systems at the different levels of a control system. Therefore it seems quite natural to apply different database structures and access methods at each level. At the lowest level the databases consist of tables which are downline loaded by higher level elements and which are updated in realtime. Part of the table contents are continuously updated or read via the communication system.

At the process control computer level, realtime database systems are required. Fast access and proper data interfaces to lower and higher level databases are required. The structures are relatively stable, which means that read and write accesses by far exceed insertions, deletions or additions; therefore access algorithms are of higher priority and can be adapted to the table structures. The table structure generation can be a kind of compilation. The amount of data often does allow memory resident solutions.

For the high level database, three features are required. Firstly, the volume of data is increasing (to more than gigabytes). Secondly, the structures must be very flexible. It must be possible to add attributes or change dependencies whilst the database is in operation. Thirdly, the database must be accessible by many users or programs at the same time. Such requirements are fulfilled by relational databases. Their only disadvantage is low speed. For this reason a careful evaluation is needed.

A proper cooperation between the databases mentioned may lead to a satisfying solution where structures are defined and changed at the highest level which also hold static data and where elements of the lower levels can be compiled and down line loaded. Saving of low level element tables into the relational database can be done triggered by events or time marks where slow responses are not important (Fig. 65).

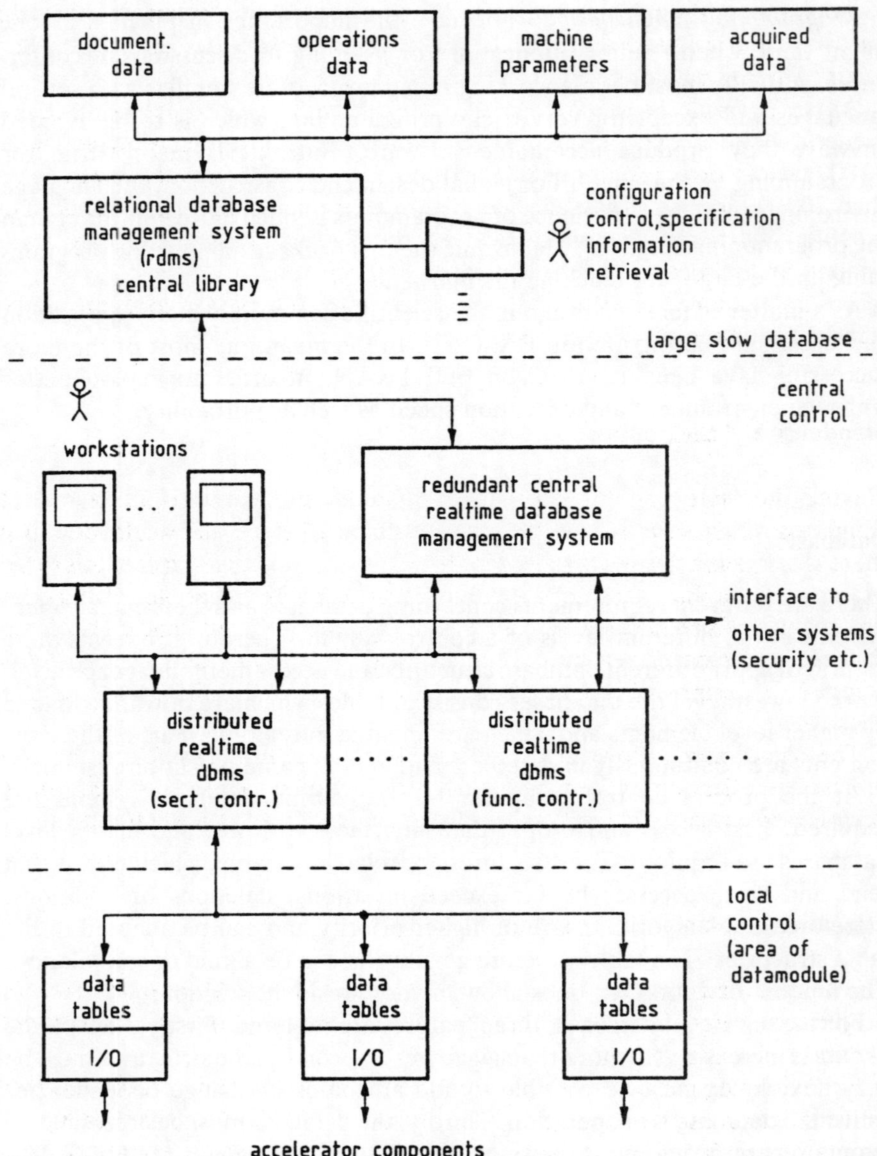

Fig. 65 *Data Organisation*

Integration of software

One of the most challenging tasks for the future of large systems is integration. There are several reasons why integration is required and why it is taking place now or will be necessary in the near future. The fast progress of control technology is forcing companies to think about stable interface ports in hardware and software and secondly large systems very often enforce vendor mixing where standards with wide acceptance are a must.

As an example of standardisation efforts the MAP project initiated by General Motors must be mentioned. Nowadays this company owns about 50 000 factory floor programmable systems and this number will grow to more than 200 000 by end of this decade. There are about fifteen large vendors in creating and testing a communication standard between all factory floor systems and minicomputer systems of the higher level.

Besides communications, languages, database systems and tools are being standardised on a worldwide scale.

Standards and their misuse

Having the huge task of integration in mind, everybody is in favour of standards which actually have been successful at all levels and worldwide. But there is also some danger that standards are misused. Let me explain this by an example. A graphics package is announced to be GKS compatible – what does this mean? Which level of GKS? Eventually we find out that calls of the package are transformed into GKS calls therefore being able to drive GKS compatible peripherals – but no program issuing GKS calls could use this package.

The complexity of software standards requires additional information like precise specification of test and validation programs especially addressing performance measures like hardware requirements and execution times.

Education and training of software producers

The desired kind of education and the training necessary highly depends on the project phase and the level within the control hierarchy. In our environment we find two types of team members.

Firstly, young people coming from schools. Here the main problem is that universities or engineering schools in general do not reach system design or programming 'in the large' nor how to evaluate products or cooperate with vendors.

Secondly experienced old fighters. They have difficulties with new technologies and like their personal style of work which may be different from a common working style based on a toolset.

My personal opinion is that conventional courses are not the right way of education and training; instead I think that small projects or well defined subprojects at an early stage of a project are the right way. Practical work has priority over theoretical courses.

Manpower and costs

Instead of a detailed analysis of these items which is not feasible in the time available let me give you some figures gathered by experience and some hints which we regarded as useful.

For the low level control systems, the close cooperation of programmers and system engineers is essential. It is efficient to put these programmers into the groups working on the subsystems and let them support all testing of the components to be controlled.

For the higher levels, teamwork is a must, using the same tools and referring to the same definitions and formats. Complex software for microcomputer systems should be written as cross software.

For the highest levels of the control software, the cooperation of application experts in the software team is desirable. These people are also well suited to be later the supervisors for an operator team.

Evaluation of software products takes at least six man months and may go up to one or two man years if it is done properly. Probably the results of such evaluations will find a wider distribution in the future. An interesting form of early evaluations are field tests, because for such tests the software is free of charge.

The cost of system software products like compilers, utilities, database systems, etc. is still relatively low compared to tools and application packages. The former are in the order of 20–50 KDM while the latter go up to several hundred KDM.

I am not addressing the problem of hiring or leasing programmers, but although this is expensive and it is difficult to integrate these team members, there is no other solution provided that the market situation does not change. It is extremely difficult to get experienced programmers.

The production of software is going to be very expensive so that the emphasis must be laid on purchasing whatever fits the requirements and concentrating on interfacing and integration.

Conclusion

Software has become a very important part of a project and the resulting technical system. Its management is a complex undertaking from the very beginning of the project far into the maintenance and development phase. Project leaders are well advised to have close contacts with software technology in order to benefit from its services and avoid the chaos caused by inefficient organisation and integration.

Literature

1 F. H. BOHN, H. HALLING: 'Experience from the Implementation of a Control- and Data Acquisition System for the Tokamak TEXTOR', Proc. of the 9th Symposium on Fusion Technology, Chicago, October 1981

2 E. R. CRICHTON, M. G. WILLEY: 'Database Management in a Distributed Process Control System', Proc. IFAC Real Time Programming 1983, Hatfield, UK

3 'The LEP Control System', LEP Design Report, CERN-LEP/84–01, June 1984

4 'Database Requirements of LEP Instrumentation', C. BOVET *et al.*: LEP Note No. 467 CERN

5 V. HATTON, G. SHERING: 'Controlling an Accelerator – The Operations Viewpoint', CERN/SPS 80–12 (Internal Report)

6 'Measurements on the Control Network of the SPS', J. ALTABER, J. P. JEANNERET: CERN/SPS/ACC 78–1

7 'Two Years of Experience with the PETRA Control System', DESY M-81/08 Internal Report

8 PAUL J. van ARSDALL *et al.*: 'Nova Laser Alignment Control System', UCRL-90033, Lawrence Livermore NL

9 P. R. McGOLDNICK: 'Supervisory Control and Diagnostics System for the Mirror Fusion Test Facility', UCRL-53090, Lawrence Livermore NL

10 G. J. SUSKI *et al.*: 'The Nova Control System: Goals, Architecture, and System Design', UCRL-86827, Lawrence Livermore NL

11 F. W. HOLLOWAY: 'Strategies of Design, Development and Activation of the Nova Control System', UCRL-89552, Proc. of the 10th Symposium on Fusion Engineering, Philadelphia 5–9, 1983

12 'SNQ Projektplan Technische Beschreibungen der Anlagenteile', Leittechnik, Dez. 1984, KFA Internal Reports for Layout of a System and Evaluation of Software

13 SNQ-Controls Overview, H. Halling, Sept. 1983

14 H. HALLING, K. D. MÜLLER, H. STOFF, W. TENTEN, K. ZWOLL, K. PÜTZ (Ingenierubüro Pütz), D. SELLGE (Biomatik, Freiburg): 'Software für SNQ-Kontrollsystem, Organisation und Strukturierung', Dez. 1983

15 W. ERVEN, S. TRENCSENI, K. ZWOLL: 'Measured Transmission Performance of an Ethernet Link', Mai 1983

16 K. PÜTZ: 'Merkmale von Datenbank-Management-Systemen, Einsatzmöglichkeiten in technisch-orientierten Projekten sowie Zeitmessungen an einem relationalen Datenbank-Management-System', Okt. 1983

17 H. STOFF, H. HEER: 'VAX Elan, Down-Line Loading and Remote Debugging über ETHERNET', Juli 1984

18 W. ERVEN, W. GRUHN: 'Concurrent CP/M86 für Microprozessoren INTEL 8086 Übersicht über das Betriebssystem. Realtime Messungen zweischen Prozessen', Okt. 1984

19 M. RUDOLPH, K. PÜTZ: 'Überlegungen und Untersuchungen zum Aufbau einer Datenbankanwendung für das SNQ-Rechnerkontrollsystem mit Hilfe des relationalen Datenbank-Management-Systems VAX-Rdb/VMS', März 1985

20 K. ANDREWS: 'VAXELN Pascal – An Outline of its Non-Standard Features', März 1985

Design of parameterizable software

José A. S. Alegria

Universidade Nova de Lisboa, Departmento de Informática, 2825 Monte da Caparica, Portugal

1. Introduction

Today's most important problem in software engineering is its poor productivity which can be viewed at two different levels. The first relates directly to the productivity of the software production process itself, which can be concretely measured, and which we all know how bad it is! The second, however, is more subtle. While a simple consequence of the first, it can only be speculated since it doesn't relate to a concrete thing but only to what society is potentially losing, in terms of productivity, for not having today a more responsive and efficient software industry. Software tools are potential productivity aids in most working activities in society. However, their quality, which in a certain way depends on software engineering productivity, has a direct influence on its user effectiveness, and productivity.

Consequently, the improvement of software productivity is clearly of extreme importance to society. A higher productivity level in software production will also trigger the development of higher quality software tools, for a more diversified set of application domains, with a corresponding dramatic impact on society's overall productivity.

2. On increasing software technology's productivity

How should we then go about substantially improving software technology's overall productivity? Well, while there are plenty of conflicting proposals around, a basic principle seems to stand out as clear as it seems obvious: current software industry is too labour intensive and a more technology intensive alternative should be found as soon as possible. So, the question above is now paraphrased: *how should we go about developing a less labour intensive and more technology intensive software technology?* While there are many valid approaches to this problem, let's continue with our top-down stepwise reasoning approach and see where it gets us.

Making things more technology oriented, from a computer scientist point of view, simply means Computer Aided something, e.g., CAD/CAM or CAE-like! That is the remedy we have advised others to take, and in many cases it seems to be working pretty well. Many, previously labour intensive industries, are quite happy with their CAD/CAM approaches. Even office work is receiving the same treatment with highly optimistic productivity improvement figures. In this arena, AI researchers are the real champions.

So, why cannot software engineering take the same remedy? The interesting thing in their case is that what they really need is a $(CAD/CAM)^2$, i.e., a CAD/CAM system for the design and manufacturing or other more specific problem domain oriented CAD/CAM-like systems. *Are current efforts in software development and support environments in this direction? Are they based on the right set of basic principles?* No! We don't think so. Software engineering environment designers should have paid a bit more attention on how our fellow partners swallowed our remedy and they would have learned something that seems obvious. What exactly? Well, let's continue and we will see it also.

First, let's observe a few basic things about, say, a CAE system. Its primary purpose is to enable the reliable design, in acceptable time, of complex circuits with the important subgoal of minimising their future real production costs, which are heavily related to structural properties of the circuits, like their board layout complexity. However, a CAE system cannot afford to restrict its attention to just those circuits properties; it must also pay particular attention to the internal symbolic representation of circuits and circuit components since that is crucial to the overall efficiency of the CAE system itself.

The inherent complexity of the operations expected from a good CAE system require far more sophisticated symbolic structures for the representation of circuits and circuit components than traditional human oriented two dimensional-drawings could possibly support. Incremental circuit drafting, for instance, require incremental and real time handling and editing of parameterised partial structures. Low level representational flexibility is here at stake. The simulation of complex circuits, being computational expensive, demands an extremely efficient representation of the circuit's operational semantics. Here, what is required is representational efficiency. Finally, for some very complex circuits, quantitative simulation is computationally intractable and AI techniques, like qualitative reasoning and simulation, are required. In this last case, representational quality from a knowledge of engineering viewpoint is what is really important. Clearly, none of these tasks would be possible, if for the sake of tradition, CAE systems remained faithful to human oriented representations, like, in this case, the old fashioned two-dimensional circuit diagrams.

The moral of the story at this point of the argument, is that the kinds of tasks a CAD system supporting the design of complex artifacts is required to so involve complex symbolic processing mechanisms, which require equally complex internal symbolic representations of the structures they manipulate, representations that are far too complex for us humans. So, a dilemma arises now. On one side, computers are capable of handling, only with advantages, far richer symbolic structures than a human is able to cope with. On the other side, humans seem to be culturally bound to particular representations which are just too ineffective as core representations in a CAD system. Fortunately, however, nothing is lost because it just happens that computers can be programmed to convey, in whatever level of detail or form a human user is fond of, those complex structures specially designed to support the CAD/CAM function efficiently.

Now let's go back to the original question of *what should the designers of most software development and support environments have learned from, say CAE designers?* Simple! Very simple, indeed! they should have learned that one must take full advantage of the ever more powerful symbolic processing capabilities of computer workstations, and of the emerging sophisticated multi-media information processing systems, to the benefit of an overall more productive software development and manipulation computer aided system. They should have learned that before an effective CAD/CAM system for software could ever be designed and built, one should first have a clear idea of the kinds of computational structures one really wants to be generating, maintaining, and analysing. Next, one should have also a clear idea of how to efficiently generate, maintain or analyse such structures, i.e., which kinds of symbolic manipulations will be required on them or on their symbolic representations. Finally, after a careful analysis of the computational requirements involved, one should then decide how to effectively represent the different structures within the CAD/CAM system. Of course, since our goal is to support the design, production and support of other problem domain oriented CAD/CAM-like systems, the problem seems recursive or reflexive, i.e., the computational structures designed and produced by the mother CAD/CAM system may themselves be the object of the manipulation and adaption by its child CAD/CAM-like system.

3. S.E. development and support environments

Then, *what is exactly wrong with current software engineering development and support environments?*

First, there is a clear confusion of what really is the purpose of a software engineering environment. There is no clear understanding of the $(CAD/CAM)^2$ view, perhaps because there is too much concern with the design, production and maintenance of embedded systems which are essential to the military and kitchen appliances. . . Fortunately, however, we start seeing healthy signs of the $(CAD/CAM)^2$ philosophy in the software industry. Of course, AI clearly leads the pack with their so called expert system shells (expert system generating environments).

Secondly, all software engineering environments I am aware of, still view the software structures they manipulate under the constraints of an artificial but 'standardised' programming language, albeit the support of sophisticated programming environments. In our opinion, programming languages as we know them today, as formal devices for the description of computations in a rigorous yet simple and readable linear sequence of characters, are completely obsolete in view of what we have said before.

This 'poetic' view of programming and programs was required when man-machine communication was limited to slow serial communication lines. Programs had to be conceived and reasoned about, off-line. They had to be written down on paper in some mutually understandable way, for both man and machine. However, the complexity of actually 'writing' the program was relatively low when compared with the complexity of conceiving it. A linguistic view of programs was then a natural, efficient and sound way out since what was at stake was simply the writing down of relatively simple and closed-ended programs and their consequent translation onto a more efficient and lower level representation (Fig. 66) by a suitable translator (**C/L**).

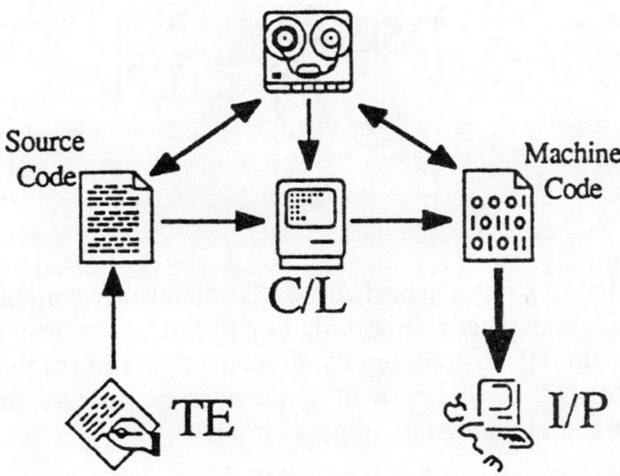

Fig. 66

Today, however, software engineers face a more complex set of problems, where the attributes of the construction process are also very important. Like in CAD/CAM systems a more graphical and interactive approach should be followed in the development of software. Here, we are no longer bound to a rigid textual language, but can directly manipulate, supported by a sophisticated multi-media interface, the internal structure of programs which is directly executable by a suitable interpreter/processor **I/Pi** (Fig. 67). We can always represent externally such structures in a variety of ways by means of special translators, be them a fancy typesetter (**ITr**) to 'beautifully' publish our programs in a commonly accepted language, or just an optimized code generator (**ICG**) to produce really efficient but unflexible versions of our structures relatively to a possibly new interpreter or processor **I/Po**.

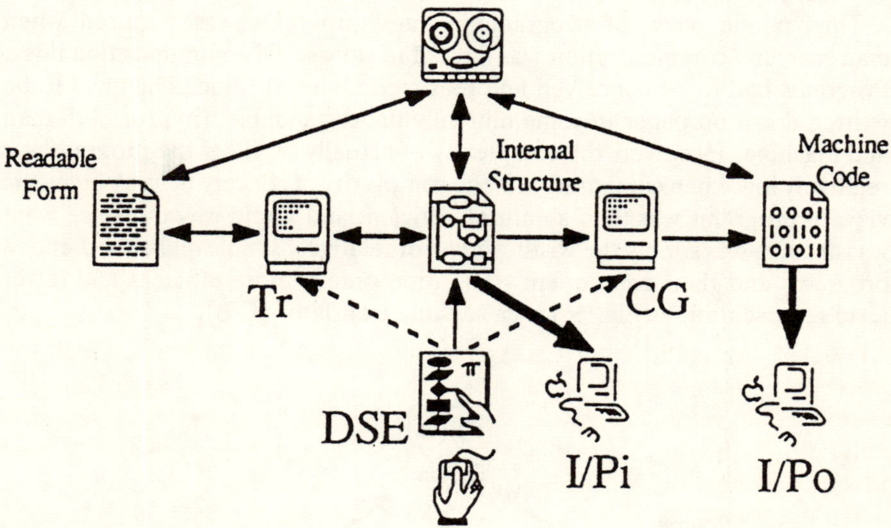

Fig. 67

However, in most cases a hybrid approach, mixing direct manipulation with text translation, is desirable (Fig. 68). For instance, it doesn't make much sense to edit arithmetic expressions by direct manipulation of their underlying tree representation. In this case it is much easier to just edit its textual representation and incrementally reparse it.

4. On representational mechanisms

Our original question of how should we go about developing a less labour intensive and more technology intensive software technology, can be reduced to an even simpler one. *Within current technology, how should we then conceptualise and represent software structures?* We don't mean the best way from the human agent or observer viewpoint. We mean the best way from a

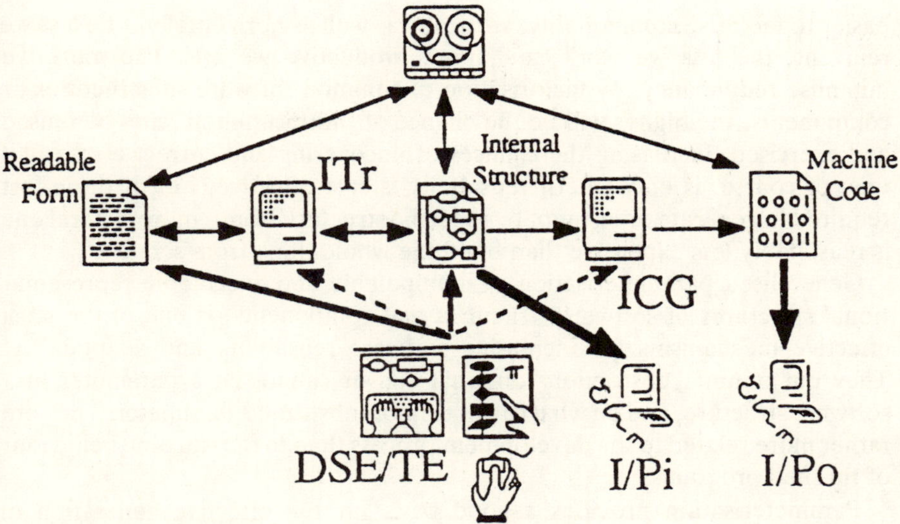

Fig. 68

CAD/CAM viewpoint, where the human perspective, as an active agent or observer, is no longer central but peripheral. It seems to us that a lot of research will have to be done before conclusive ideas are reached. Unfortunately, this issue has received little focused attention by researchers in Software Engineering. However, a few things seem already clear to us.

First and foremost, as we've argued, we think that persisting on viewing programming as formal text writing, albeit supported by sophisticated environments, is a serious mistake and should be abandoned. If we are to work towards a truly flexible and productive software manufacturing technology and industry, we must start concentrating on effective means of symbolically representing software structure. Efficient symbolic representation mechanisms for software components viewing programs not only for what they stand for (i.e., for what they mean and do when interpreted by a suitable interpreter), but also, and ever more importantly, for what they are as 'real' computational structures. These representations should simplify the rapid construction of prototypes, displaying enough useful structure to enable an aposteriori 'adaptation' to acceptable production versions satisfying stricter resource requirements. These production versions, in turn, should remain sufficiently flexible to allow further user level incremental adaptations.

Secondly, we believe that in addition to a different view of software representations, we should determine what to do exactly with them in order to maximise productivity and reliability without loosing flexibility. Clearly, reusability is a sound basis for a more capital intensive approach. So, software structures shouldn't be just easy manipulable computational objects. They should also enable the minimisation of structural redundancy, by making it easier to factorise commonality, vertically as well as horizontally. The less we reinvent, the less we work, and more productive we are. The more we minimise redundancy, by factorisation of common software substructures or components, the higher will be the chance of those common parts be reused and exercised, increasing the chances of uncovering and correct errors. All this, of course, is only true if reusability is cost effective, i.e., if the effort required to generate or instantiate a specific structure from a more general one is reasonably less expensive than build the whole thing from scratch.

Generalised parameterisation of manipulable and executable representational structures of software structures and components, is one of the most effective mechanisms for increasing software reusability and adaptability. They put minimal restrictions on what can or cannot be a parameter in a software structure, and on what can or cannot substitute a parameter. They are rather more related to the development process than to the run time behaviour of finished programs.

Parameterisation provides a good basis for the effective generation of alternative concrete versions of a software product line, out of a common and highly parameterised abstract software structure template. At a higher level, and relative to the template they were generated from, each such alternative behaves essentially the same. At a more specific level, however, each of them may differ substantially in the way specific behaviour is concerned, particularly in the way restrictions and requirements are put on resources, including the user, the most important resource of all.

The keywords are structural factorisation and abstraction. Basically, a generalised parameter of a software structure is any substructure which makes sense to factor out and abstract into a 'virtual' component. A generalised parameter is simply a component substructure, of a software structure, that can be statically or dynamically expanded, replaced, or simply 'edited' into a 'virtually' equivalent alternative, i.e., one which satisfies the same 'virtual' interface protocol. Its scope and power, however, are intimate with our view of software structures, particularly the representation we choose for them. We rely on a vertical/horizontal conceptual taxonomy to keep explicit structural factorisation.

If we consider, for example, the simplistic view of software structures as represented by parse trees, a non-terminal would be the closest analogy to what we mean by a generalised parameter. A non-terminal can be expanded or edited into any of its derivation subtrees, even one which is partial, i.e., one with further non-terminals in it. Likewise, a particular derivation subtree can be collapsed or parsed back into its generating non-terminal or, for that matter, into any of its intermediate generating derivation subtrees. Different degrees of computational complexity are involved, yet most current structure oriented syntactic editors do exactly some of these operations. The analogy, however, stops here. Our factorisation mechanisms have a clear semantic counter-part.

Truly generalised parameterisation, however, must go beyond the process of flexible software generation; it must provide a safe ground for incremental, user driven, adaptability of software. A concrete instance of the generation/instantiation process, should be nothing more than a parameterised and adaptable core, tailorable by the user to his particular needs, by his hand, and when he do desires. Not just cosmetic changes but extension or specialisation of specific functionality.

Designing software for reusability encompasses the design of generic cores which are flexible enough to be incrementally and safely adaptable to specific user requirements and by himself. the already popular spreadsheet is an approximate example of this trend.

However, the full power of incremental parameterization is intimate with the software structure itself. Traditional 'poetic' views of software, as pieces of text in a given strict programming language, do not provide enough structure to be factorised and abstracted from, and in that way giving place to potential places for controlled change. AI software structures, on the other end, are the product of languages where self-manipulation is a common mechanism, where program parts are directly manipulable structures and therefore give more freedom and opportunity to establish places for change.

5. About the talk

After all this long deflamatory discourse are we going to give you the final answers to the remaining questions? Of course not. We have just pointed at what from our viewpoint seems wrong. That doesn't mean that we have a definite alternative to what we feel is wrong. We will simply illustrate a possible way of viewing software structures where generalized parameterisation is taken seriously. The talk will illustrate a particular representation approach of software structures that in some restricted sense answer all the above questions.

6. The example

The basis of our commented illustrations will be a spreadsheet program (Fig. 69). We find it an excellent example for the following reasons:

- It is the most popular example of a problem specific environment[1] with a clear goal of improving the productivity of those working in a rather large class of problems which can be represented and manipulated within the general framework of mathematical spreadsheets.

[1] A good approximation of what we mean by a CAD/CAM-like problem domain oriented working environment.

DataStructures.Grades

	A	B	C	D	E	F
1	Question -->	Question I	Question II	Question III	Total	
2	% of total -->	40	30	30	100	
3						
4	Student	Question 1	Question 2	Question 3	Final	Warnings
5	Name	% points	% points	% points	Grade	
6	Ann Mary	75	50	65	65	
7	Charles Gregory	75	65	40	62	BEWARE!
8	Joe Bloom	50	75	75	65	
9	John Smith	100	100	95	99	
10	Joseph Lamas	75	75	50	68	
11	Mary Wilson	90	75	75	81	
12	Peter Charles	50	45	45	47	FLANKED
13						
14	Total	515	485	445	487	
15	Min	50	45	40	47	
16	Max	100	100	95	99	
17	Avg.	73.57	69.29	63.57	69.57	
18	Std. Dev.	18.64	18.35	19.73	16.37	

Fig. 69

- It is a good example of the hybrid (Fig. 68) approach of dealing with software structures. A large part of the user work with a spreadsheet is reduced to mouse clicks, selecting different menu options, selecting cell groups, etc. The reference of cells, as part of arithmetic expressions, can be done by pointing at them with the mouse, or, alternatively, by simply typing in their name.

- It is an excellent basis for discussing parameterization mechanisms. A spreadsheet may involve the following levels of parameter types:

  - Potential parameters that the implementer prefers not to bother with and rely instead on an underlying general purpose mechanism. For example, the actual number of spreadsheet cells can be left open, and if given can be taken as a simple guideline to improve the underlying's memory management efficiency.
  - Parameters which the implementer has restricted to a closed set of well defined alternatives. There is a constant default value which the user may set to his own taste. The user cannot, however, alter or extend the set of possible alternatives. The default displaying style of the spreadsheet exemplifies such type of parameter.

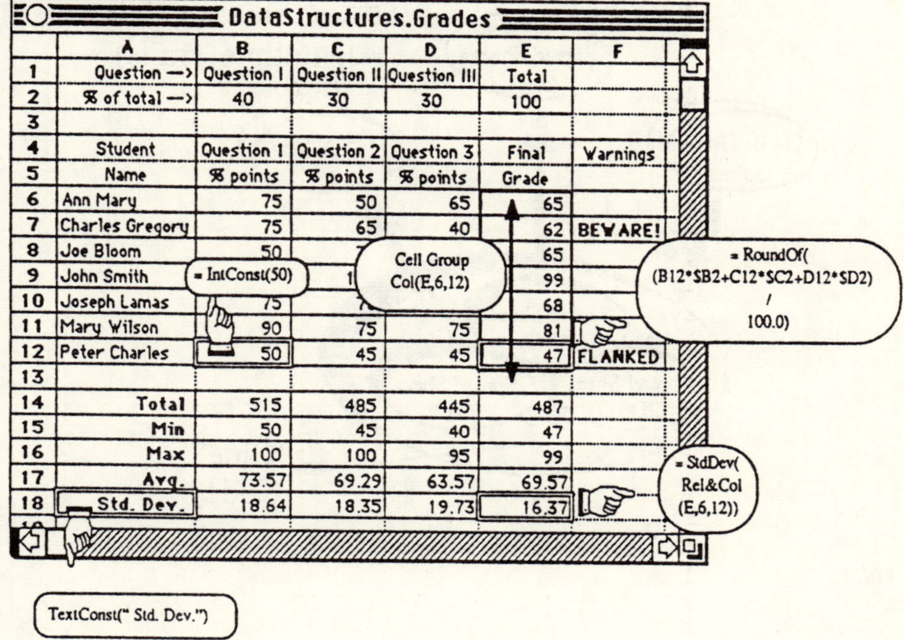

	A	B	C	D	E	F
1	Question —>	Question I	Question II	Question III	Total	
2	% of total —>	40	30	30	100	
3						
4	Student	Question 1	Question 2	Question 3	Final	Warnings
5	Name	% points	% points	% points	Grade	
6	Ann Mary	75	50	65	65	
7	Charles Gregory	75	65	40	62	BEWARE!
8	Joe Bloom	50	7		65	
9	John Smith		1		99	
10	Joseph Lamas	75	7		68	
11	Mary Wilson	90	75	75	81	
12	Peter Charles	50	45	45	47	FLANKED
13						
14	Total	515	485	445	487	
15	Min	50	45	40	47	
16	Max	100	100	95	99	
17	Avg.	73.57	69.29	63.57	69.57	
18	Std. Dev.	18.64	18.35	19.73	16.37	

Fig. 70

- Parameters which the implementer has restricted to an initial open set of alternatives, but that the user is free to edit to his own needs, including the default value. The default format depends on the cell's content type.
- Parameters whose default value is a function and not a constant. The default format of a cell depends on the cell's content type.
- Parameters whose value is established at the time the spreadsheet system is manufactured. The internal concrete representation choice for the spreadsheet is one of these cases. This aspect is presented in the commented illustrations later.
- Parameters whose set of alternative values can be changed dynamically by the user himself, but which involve the adaptation of the spreadsheet computing behaviour. An example of this last case is the cell contents (Fig. 70). One can put expressions which envolve an open-ended set of possible operators. A user may want to define new operators to his own needs. Or, a developer may want to sell different spreadsheets with different 'built-in' sets of operators. This aspect will also be discussed in detail during the talk.
- Parameters which deal with computing strategies, like the order in which the spreadsheet cells should be evaluated.

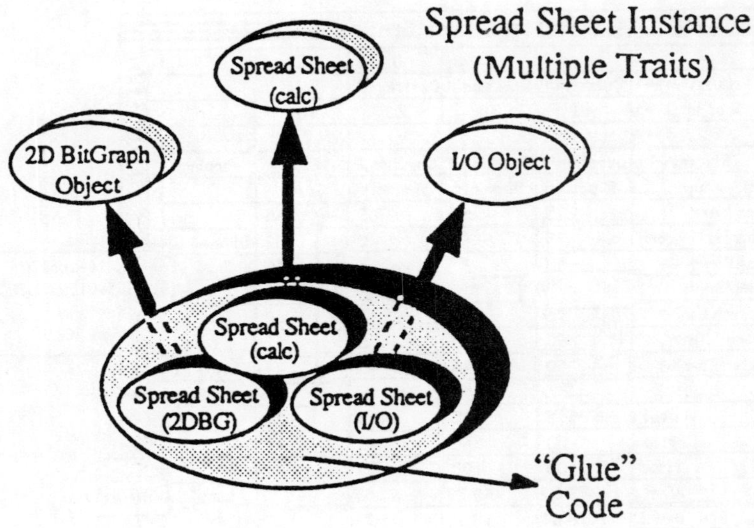

Fig. 71

– Finally, a spreadsheet is a good example of how orthogonal facets are put together in a complete and coherent product. A spreadsheet has at least three different and orthogonal facets (Fig. 71):

- A calculation facet which embodies the software structures responsible for actually doing the calculations. This could have to do with a generic two-dimensional sparse array of cells containing pointers to an open-ended set of expressions, represented in some suitable form. In addition an evaluation strategy would have to be defined.
- A graphical facet embodying the structures responsible for the graphical representation and manipulation of the spreadsheet.
- An I/O facet which would be responsible for dealing with the I/O aspects to and from a file system.

Software development productivity would be greatly improved if one could factorise software structures in horizontal facets, each with an orthogonal facet, placing minimal interference on each other. Horizontal factorisation will not be discussed in any great detail. We will concentrate more on vertical factorisation in the commented illustrations that form the rest of this paper.

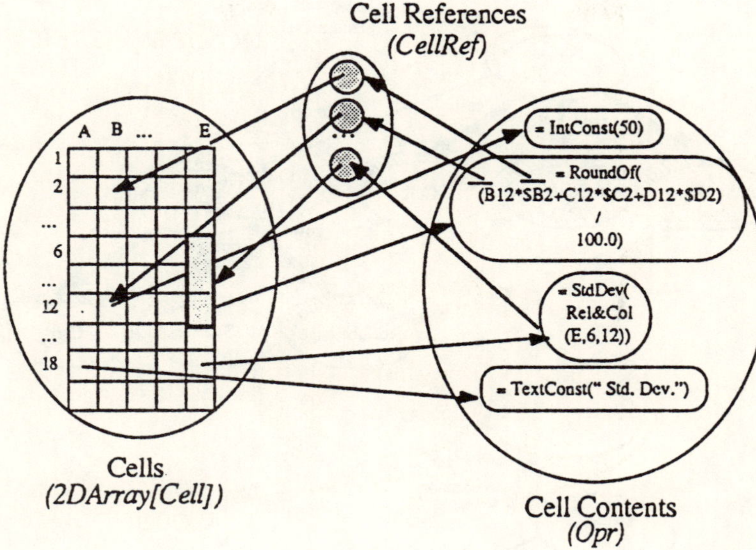

Cell References
(CellRef)
A B ... E
1
2
...
6
...
12
...
18
= IntConst(50)
= RoundOf(
(B12*$B2+C12*$C2+D12*$D2)
/
100.0)
= StdDev(
Rel&Col
(E,6,12))
= TextConst(" Std. Dev.")
Cells
(2DArray[Cell])
Cell Contents
(Opr)

Fig. 72

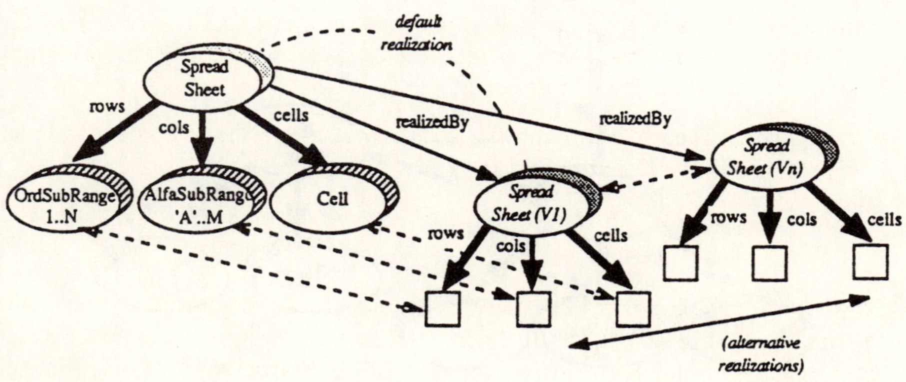

default
realization
Spread
Sheet
rows
cols
cells
realizedBy
realizedBy
OrdSubRange
1..N
AlfaSubRange
'A'..M
Cell
Spread
Sheet (V1)
Spread
Sheet (Vn)
rows
cols
cells
rows
cols
cells
(alternative
realizations)

Fig. 73

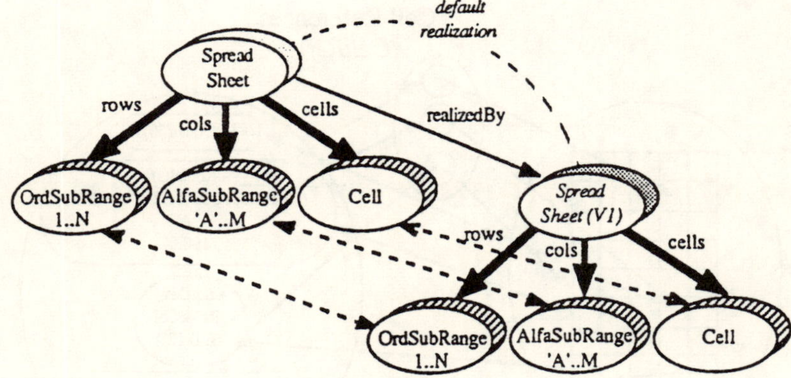

Fig. 74

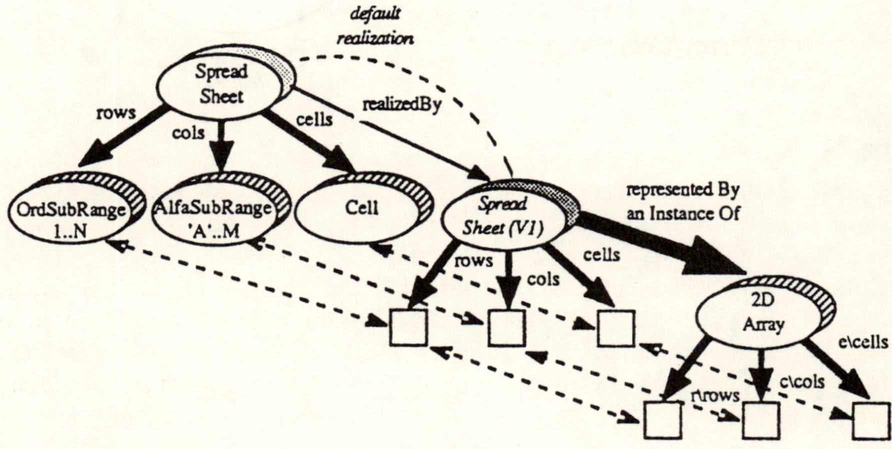

Fig. 75

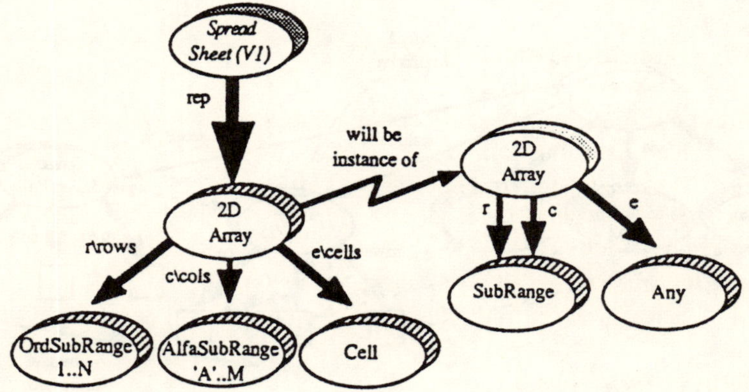

Fig. 76

An object-oriented database system: Object-Base

Herbert Weber
University of Dortmund

Silke Seehusen
University of Bremen

Abstract

An object-oriented database system is a database system in which the concept of abstract data types is used strictly for describing and using the database. In consequence the data are described in conjunction with the appropriate operations, the only operations allowed on that data. The object-oriented database system Object-Base, presented here, allows a hierarchy of data types. That kind of database description enables the system to be adjustable to very different application characteristics. Different levels of user interfaces, from the unexperienced casual user to the sophisticated system programmer, are provided.

Table of contents

1. Introduction

There are many database management systems (DBMS) available now, and many of them are really good ones. Each of these DBMS is supporting one of the different standard data models (hierarchical, network, relational). Many additional tools, e.g. data dictionary, are provided for managing the databases.

Most DBMS are designed for an administrative or commercial environment. For that application area they are universal. The same DBMS may be used for a university's library and for an inventory of an automobile producer. But the university has the disadvantage that the DBMS cannot be tailored to each application environment. Such adjustifiability is important with respect to e.g. minimise the effort for application programming on the one hand and with respect to tuning the whole system on the other.

We want to cope with these problems when designing an object-oriented DBMS. Object-orientation has become an important concept in software development techniques.

The concept can be considered only in coherence with data abstraction [Liskov 75] and modularisation. Thus it enhances comprehensibility, correctness and maintainability of system design and implementation. It has influenced newer programming languages (e.g. ADA; Modula-2, see [Wirth 82]) and operating systems. We want to make use of this concept in database systems as well.

2. Why yet another database system?

Some deficiencies of contemporary database systems and the following new requirements motivate the newly developed database system.

For in-situ database system support on workstation computers
Many existing commercial applications and many new applications of database systems seem to need in-situ database services provided at workstation computers. These computers provide limited memory and processing capacity that may not suffice to support generalised database management systems of the kind described above.

For application adjustable database management
More importantly, however, the existing and the new database applications on these workstations differ considerably and are thus inadequately supported by only one standardised or generalised type of database system. They are, however, not that different to justify a completely new database system concept for each new application. It, therefore, seems to be attractive to have a database system that can be easily adjusted to many applications with the adjustment much less expensive than a complete redevelopment.

The new type of database system that is going to be described here is meant to be that kind of an adjustable database system serving for a variety of applications on multifunctional workstation computers.

For different data model support
In-situ database management facilities on workstation computers are different from the more generalised database management facilities in a number of other ways. First of all, they serve as a tool in a narrow application environment with rather specific requirements for data representation and manipulation. In those environments data representation in a form most suitable to the specific application is not only a wish but a very strong demand in order to fulfil the in-situ service requirements. This additional requirement does not only ask for the support of different data representations e.g. of different data models for different application environments. Very frequently the support of different data models (i.e. different views of data) in the same application environment is a must. As a consequence the database system for workstation computers should be adjustable to support different data models in different installations of the system but also in one installation simultaneously.

For semantic data model support
Since the data models supported by the database system are supposed to meet the requirements of particular applications they are usually expected to be semantically richer than, say, the relational model of data. For the support of specific applications semantic data models that enable the representation of static (i.e. declarative) facts and of dynamic (i.e. behavioural) facts are needed to be supported by the system. The system described in here is built to support in that respect.

For complex object representation and manipulation
Since the system supports a powerful abstraction mechanism arbitrary complex objects for specific applications may be defined and represented in the database. In addition, the management of the workspace in main memory for complex objects may be simplified with tailored query facilities that do not fetch entire complex objects from secondary memory but allow a selective retrieval of components that are needed at the time. This is not to claim that all complex object representation and manipulation problems for all types of applications are automatically solved with the system. But the claim is made that Object-Base provides all the facilities to adjust the system to different kinds of complex object problems in a flexible manner.

For long transaction management
Some more uncommon database applications require the execution of transactions of a very long duration (maybe hours or days). The conventional

transaction management schema is not adequate to handle very long transactions. Object-Base treats every transaction as a nested transaction if the database objects it accesses are composed in a hierarchical fashion. Therefore, the complex object management along with the nested transaction management schema are the prerequisites for the management of long transactions. In addition, transactions only serve as a unit of consistency. Recovery from conflicts and failures in the execution of concurrent transactions are handled with a different schema that does not 'undo' partially completed transactions but computes 'compensational transactions' on data base objects affected by not fully completed transactions.

For truly casual user support

In addition a new class of users comes into existence with the advent of workstation computers in new application areas. They are primarily interested in the use of prefabricated application programs for a number of reasons:

- They do not have the knowledge and skills to develop their own application programs, or
- they cannot find qualified personnel for the development of application programs, or
- they cannot afford to employ the rather expensive program development personnel, or
- they cannot spend the time and effort to get themselves educated in the program development profession, and finally
- they are not interested to get themselves involved in the program development but think of it as a bought in service.

The users of the new in-situ database systems are also to an ever increasing extent of that nature. They want to run their application programs but cannot involve themselves for the many reasons mentioned above into the development of their own application programs. It was assumed for a while that modern high level – non-procedural – query languages like SQL would allow so-called casual users to use a database system. The group of users, however, as has been characterised above, cannot or will not be able to handle a high level programming language that hosts a particular query language and/or the query language itself to write its application programs.

Instead a program generation facility seems to support these users in a much better way than general query facilities. The program generation shall be accomplished through the synthesis of application programs from prefabricated primitive building blocks in much the same way as program generators built programs from prefabricated macros. The group of users mentioned above is meant to be supported by the Object-Base and its associated application program generator as described later in the paper.

For low system overhead
Last, but not least, many of the conventional applications of databases are very static in nature. They are primarily used for the same kind of standard applications over and over again. Additions to the established applications seldomly occur, existing applications almost never disappear.

The use of a database system that supports precompilation and interpretation of queries is not justifiable for this kind of application pattern. The existence of both increases the complexity of the database management system considerably, increases the general overhead and leads consequently to a reduced overall performance. Systems that only provide a precompilation facility are expected to suffice in many cases. The Object-Base is aimed at being used for the stable kind of applications mentioned above giving its user a system of lower complexity and lower system overhead. Because of its adjustable nature the system may, however, be easily enhanced by an interpreter facility later when needed.

For general purpose database management
Although Object-Base has been developed especially to serve as an in-situ database tool it may of course also be used as a general purpose database system that serves as a central data repository for a large community of different types of users. It can serve in that respect if a flexible interpretative query facility is not needed and the application environment only requires infrequent changes of application programs, integrity assertions, access constraints etc., or if an interpreter facility is added on later.

3. Object-Base: A database and application-generator system for multifunctional workstations

Powerful workstation computers supporting multi-user operating systems like UNIX are increasingly used in many applications areas like office automation, CAD/CAM, software development etc. All these applications need to be supported by adequate data management capabilities. Since the applications are rather different in nature, different database support functions are needed for them. The Object-Base is aimed at being an adjustable database system for multifunctional workstations that can be tailored efficiently towards different applications.

Workstation computers are also increasingly used by non-sophisticated programmers but rather by end-users. They are primarily interested in running prefabricated standard application programs. Database accesses are made possible for this type of user through the execution of predefined standard queries.

Object-Base provides means to flexibly combine predefined simple queries into complex constructed queries thus enabling the generation of application programs from preformulated 'query macros'. This feature of Object-Base is called Application Program Generator.

Like any other relational database system Object-Base may provide relational standard operations (e.g. SELECT, INSERT) on standard data types (e.g. RELATION). These may be used by the application programs and by sophisticated users.

Each user interface is provided by one or more modules, the object descriptions. An overview of a very simple database is given in Fig. 77.

Every end-user has his/her own view of the database, his/her external interface she/he is interested in. The external interface provides the end-user with operations she/he may execute with appropriate parameters.

The external interfaces are thus different to external schematas [ANSI 75] of a conventional DBS. An external schema is a view of data, and only standard operations (e.g. SELECT, INSERT) are allowed on the data. An external interface is not meant as an interface for an application program. It is designated to the end-user who rarely calls the operations of his/her external interface within a program.

The description of the data and operations of the database is performed in a modular fashion. The hierarchy of application modules plays the role of a conceptual schema of the database although the operations, normally specified within application programs, are integrated. In Fig. 77 a simple module hierarchy is depicted consisting of four modules only.

The application modules are based on the so called standard modules provided by the underlying database management system. They include basic operations and data types useful for specifying the application modules.

The different user interfaces are discussed in detail in the following sections.

3.1. End-User interface

Object-Base is meant to support the truly casual user who does not have an interest in programming database accesses in a higher level query language (e.g. SQL), but whose interest is truly in the execution of existing application programs. He uses the database system in a 'push-button fashion'. For that purpose the user will be provided with a structured menu that allows him to select the application program he is interested in. He starts its execution by prompting the identified application program. This can be nicely supported with a mouse and a screen editor if available. For this characteristic the system is also called a 'push-button-system'.

Each end-user gets his/her own external schema that identifies all his/her private interface entities he/she may want to make use of in his/her access to the data base. The entities identified in the external schema are operation names that identify operations on types of data objects visible for that user.

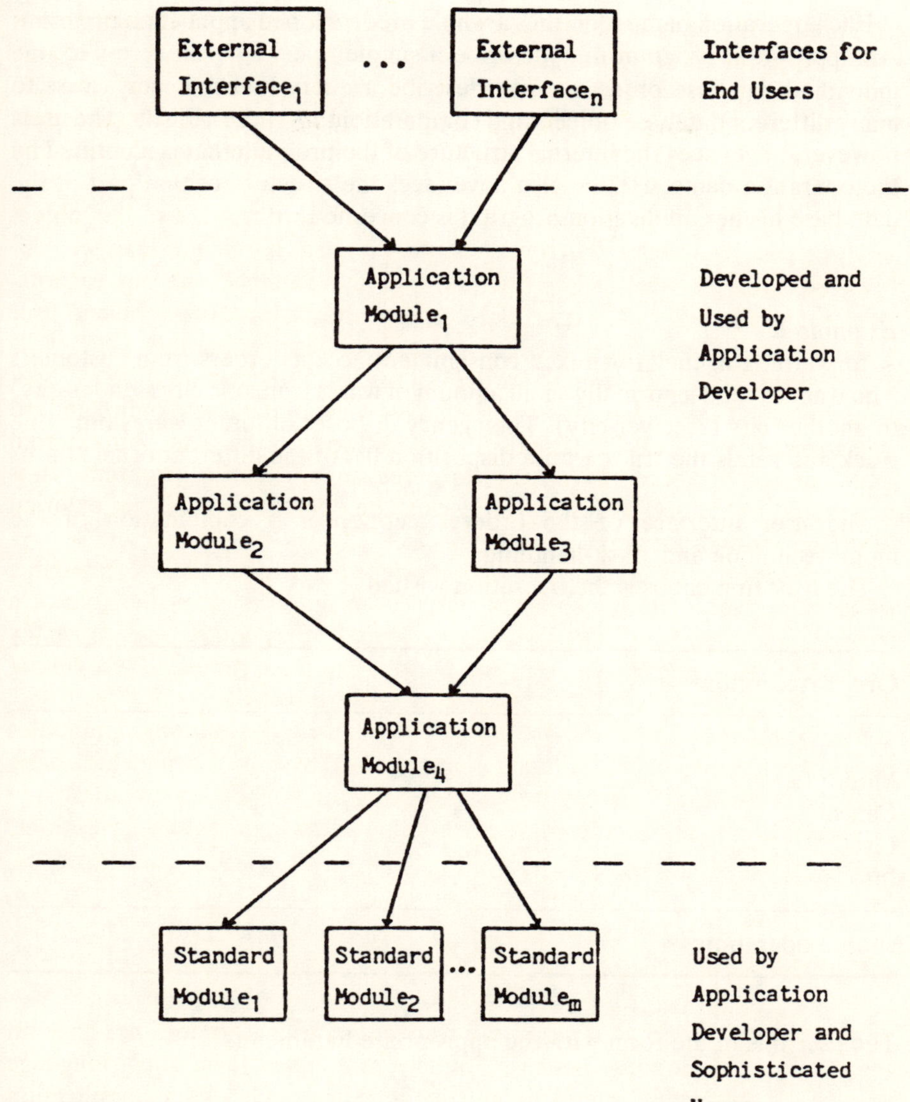

Fig. 77 *Module hierarchy of a simple database*

The operation names are associated with placeholders for parameters the user is asked to supply prior to the execution of the operations. After all parameters are supplied properly the user may prompt the execution of an operation.

Each operation name identifies a whole predeveloped application program. An application program may describe a simple query to seek access to one individual database object or may describe a query that involves access to many different database objects and computations on those objects. The user, however, never sees the internal structure of the program that is identified by the operation name. He/she also never sees the structure of that part of the data base his/her application program is concerned with.

Example 1:
A forwarding agency for mixed consignment accepts orders from customers who want to send a normally small amount of wares from one city (sender-city) to another city (receiver-city). The agency disposes different wares onto one truck and sends the truck with a disposition list of the different orders on its way.

The user interface of the order acception is a combination of the menu-technique and mask-handling.

The user first chooses the operation wished

Order Acception

Accept
Modify
Cancel
List
Info

---

choose operation:

The user fills in the form with the appropriate Parameters

Order Acception: Accept

Order No.:	(No. or N for new number)
Customer (Sender):	(Name and Address)
Receiver:	(,, ,, ,,)
Weight: . . . kg	(kg)
Special Issues:	(Text)

---

fill in forms

The system provides automatic integrity inforcement. The execution of an application program will only be completed if all integrity constraints that have been declared for data in the database will not be violated during execution of the application program. The system will recognise integrity constraint violations and notify the user of the abortion of the execution of this application program or prompt him/her to either change wrong parameters he/she has supplied to the program or to supply additional parameters etc. The system, thus, guides the user in the work with the system.

The system also allows only accesses to the database that are legal for the respective user. The system automatically checks on the access privileges each user has been granted.

3.2. Application Programming

Object-Base supports the construction of the database for a specific application and the associated application programs that embed accesses to the data base. This task can only be accomplished by professionals with a higher degree of knowledge on programming and database management. They still do not need to be sophisticated programming and database experts. Object-Base will support the definition of the database and of application programs to the extent that users of the system in that mode may concentrate on the analysis of the application and neglect details of programming and data base design. This will be made possible through a generator facility that enables the simple construction of database schemata and application programs. For that purpose the user will be provided with a system supported by very high level description and structuring discipline for elementary building blocks of databases and application programs. For this characteristic the system is also called an 'application generator'.

In this mode the system provides support for the construction of the database and of application programs from building blocks through the definition of new building blocks and the re-use of prefabricated building blocks, the standard modules and the already defined application modules. It also supports the formulation of integrity constraints and access constraints in an unambiguous fashion. A syntax-directed editing facility that is not an integral part of the Object-Base may be employed to partially automate the development and validation of the building blocks and their proper interconnection.

The user gets access to an extended conceptual schema that identifies all entities he/she may want to make use of in the synthesis of application programs. The entities identified in the extended conceptual schema are module descriptions [Weber 83].

A module description consists of

(I) the description of one type of data object,
(II) the description of all operations that may be applied to the kind of data object described in the module (i.e. prefabricated simple queries).

Example 2:
The orders of the forwarding agency introduced in example 1 are specified. This is a simple version of the module 'orders' described in pseudo-code. The module 'orders' uses the module 'order' managing one order.

```
module orders
module interface
purpose: managing orders
permanent data: ORDERS
operations: accept (in  sender_name, from_city, receiver_name,
                        to_city: NAME, weight: KG, charge: DM,
                    out No#: NAT, charge: DM, message: MESSAGE),
            cancel (in  No#: NAT,
                    out  message: MESSAGE)
module body
permanent data ORDERS = set of ORDER
operation accept (in  sender_name, from_city, receiver_name,
                      to_city: NAME, weight: KG) charge: DM
                  out No#: NAT, message: MESSAGE)
    call order_numbers. new_number (out No#);
    call order. create (in  No#, sender_name, from_city,
                            receiver_name, to_city, weight, charge,
                        out    message)
end accept

operation cancel (in No#:NAT, out message: MESSAGE) is
    call order. delete (in No#, out message)
end cancel
end module orders
```

Application programs will be synthesised from pre-fabricated simple queries on data objects that have been described in the respective module description.

All the modules the simple queries are taken from to formulate an application program are said to be the constituent modules of that application program.

A newly formulated application program will be made a constituent part of a new (i.e. higher level) module with the composition of data objects of all constituent modules as its new data object. This features enables the hierarchic composition of application programs from pre-fabricated building blocks. We call the newly formulated module that hosts an application program an application program module (in short: application modules see Fig. 77). After its formulation it becomes a pre-fabricated module for later use in other even higher level application program constructions.

Example 3:
Assume the forwarding agency of the proceeding examples is already managing a database for orders and one for freight charging including the different tables for freight rates. The end-user uses these two databases separately. When accepting an order s/he first calls the freight charging program to charge the freight and then fills in the amount in the mask of the order acception.

There are the modules:

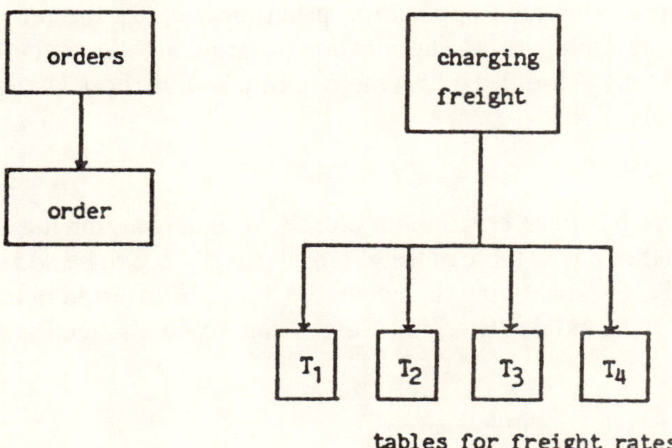

If now the freight has to be charged automatically when inserting an order the interface of the modules 'orders' and 'charging freight' are used as part of the abstract level language provided by the already existing module hierarchy:

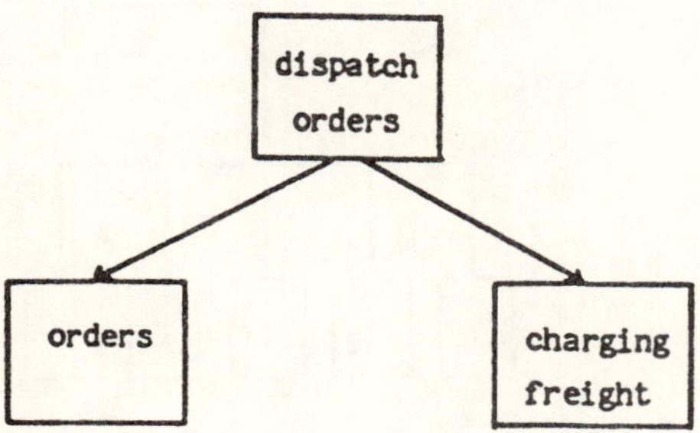

The operation of dispatching an order would look like:

```
operation accept (in:  sender_name, from_city, receiver_name, to_city:
                       NAME, weight: DM
                  out: No#: NAT, charge: DM, message: MESSAGE)
                  is
      call charging_freight.charge_freight
                  (in from_city, to_city, weight,
                  out charge, message)
        if message = ok then
   call orders.accept (in  sender_name, from_city,
                           receiver_name, to_city,
                           weight, charge,
                      out  No#, message)
        fi
end accept
```

Since a new service is provided the end-user interface may be enriched now.

Changes in pre-fabricated application programs require the recompilation of the constituent modules of the application program module that are affected by the changes only. Modules that remain untouched by those changes do not need to be recompiled.

Example 4:

If no new service has to be created but e.g. the freight charging itself has to be changed because of a new table of freight rates, say T_5, has to be added. Then a new module, T_5, encapsulating the new table has to be inserted in the module hierarchy and the module 'charging freight' has to be changed accordingly.

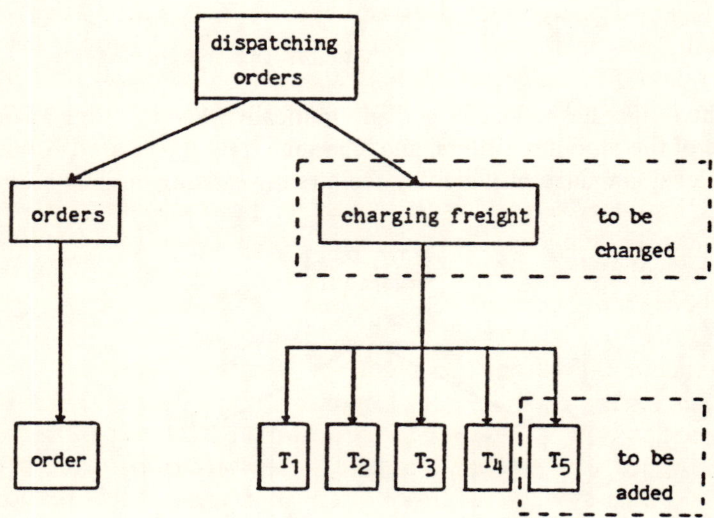

Because the interface of 'charging freight' remains unchanged no other module has to be changed. An end-user would not notice this kind of maintenance.

Application programs may also be constructed through different additional combinations of operations defined in pre-fabricated modules. To enable application program synthesis through additional combinations of operations all modules that participate in the new combination are needed to be re-compiled. All unchanged modules do not need to be re-compiled.

New application programs may also be built through the addition of a module to the database and through the replacement of a module already existing in the database system. New modules can be added by only compiling the new module and linking it appropriately to the set of existing modules. The replacement of modules also requires the compilation of the new module and the proper decoupling of the ancient module as well as the proper linking of the new module.

3.3. Basic interface: relational query interface

In addition to the two interfaces mentioned above a third one is provided that offers a truly relational query language for the retrieval and change of data. The interface is somewhat subordinate to the other interfaces in the sense that it is meant to provide access to the basic data base objects. All user-oriented types of data visible in the external schemas or in the conceptual schema are ultimately represented by unnormalised relations. Anomalies in change operations will be avoided by tailoring operations to each individual relation rather than by decomposition programs by using appropriate primitives of the relational query language. Integrity constraints and access constraints, however, may only be defined for individual base relations. Object-Base used in this mode inforces only these kinds of constraints and not interrelational constraints. It therefore provides only in part what is now frequently termed 'integrity subsystem'.

Using this interface Objects-Base acts like a conventional relational database system. The database appears as a collection of base relations and queries may be formulated in a relational algebra type query language. The use of the system in this mode is reserved for sophisticated data base and programming professionals.

The system allows the definition of integrity constraints and access constraints for the database and of application into Normal-Form Relations and the application of universal query operations. The user of the relational query interface will, however, not be aware of the existence of tailored operations. S/he uses the relational query language in its conventional way.

The Object-Base is capable of selecting the appropriately tailored operation on the basis of scope information supplied with the relational query.

The third user-interface is not necessarily restricted to offer relational query processing. Other data models may be supported instead or simultaneously with the relational data model. This is made possible by the earlier described module replacement capability of the system and by the encapsulation capability of each type of data base object mentioned before.

4. General architecture of Object-Base

Object-Base is a multi-user system that allows a number of users to concurrently use the system in different usage modes (see chapter 3). The system is entirely structured of separately compilable modules. The modules exhibit a data encapsulation property that enables a rather flexible reconfiguration of the system through additions, removals and replacements of the modules in the system structure.

An overview of the Object-Base is depicted in Fig. 78. The main components are described in detail later on in this chapter.

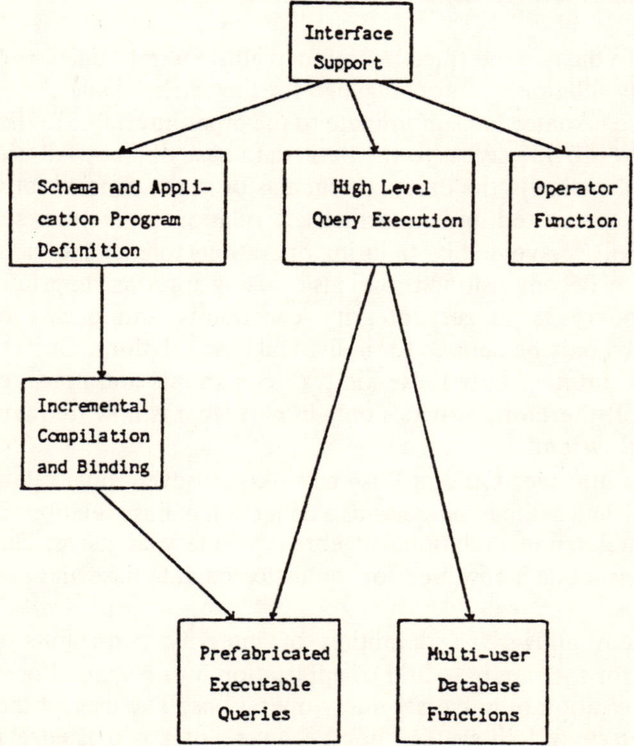

Fig. 78. *Gross Object-Base system architecture*

The only access to Object-Base is via the Interface Support. It provides the different user interfaces described above. Additionally but nevertheless necessary is the interface for the operator. The Interface Support checks the identity of a user and gives her/him access to only those operations s/he is allowed to.

The Operator Functions include the operations for starting, stopping and initiating maintenance operations, e.g. recovery operations. All these operations are necessary for running the database system.

The Schema and Application Program Definition supports the tasks of database administration and application programming. Facilities for defining, adding, deleting and changing modules are provided. New and changed modules are compiled and connected to the other modules by the Incremental Compilation and Binding. The executable code of the modules is managed by the component called Prefabricated Executable Queries encompassing all programs of executable queries of the database.

When a user initiates via the Interface Support the execution of a query this query is executed by the High Level Query Execution. Only this component has access to the database via the Multi-User Database Functions. The database functions encompass all functions accessing the database, which are necessary in a conventional database system as well.

As outlined before the facilities provided by the system differ from conventional system concepts. The differences will be described below.

The general notation used here to describe and depict the system functions is as follows: A system function is defined as a processing element with its associated input and output and in addition with the state data that are subject to changes during the execution of the processing element. This may be depicted as follows:

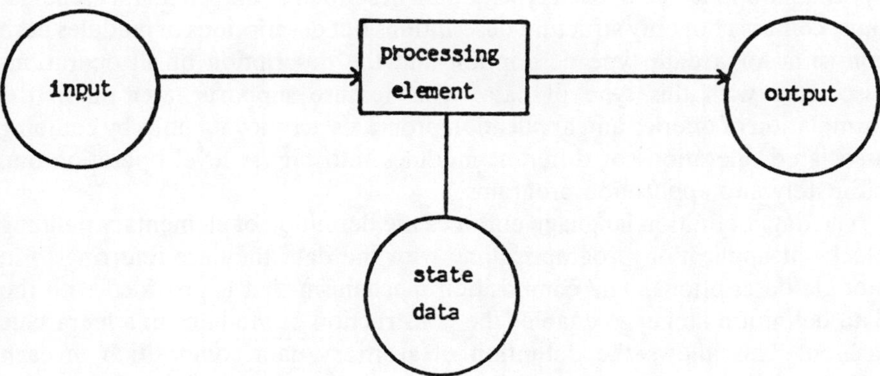

We may look at the data base system itself as an example for the application of this notation.

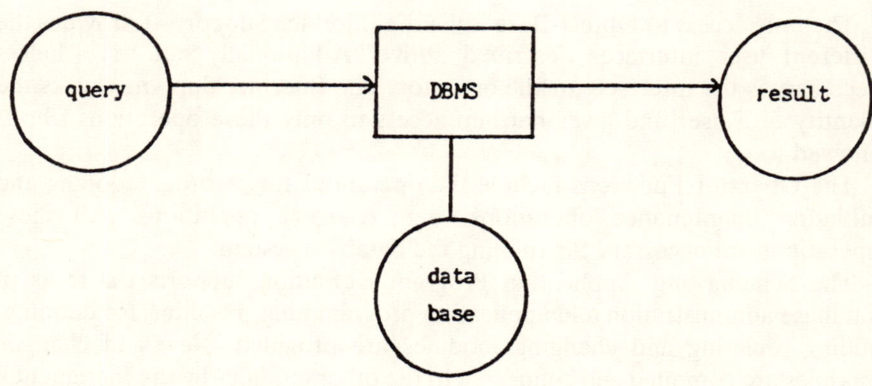

The processing element 'DBMS' takes a query as an input and produces a result and may change during its execution the state data called 'data base'.

This description schema will be used to describe the main functions of Object-Base.

4.1. Schema definition and application program generation

Schemas serve in Object-Base as in other systems as data descriptions. In addition to data descriptions schemas also encompass descriptions of all operations applicable to each type of data described in the schema. A schema, thus, contains not only structure descriptions but descriptions of modules each consisting of a data type description and the description of all operations associated with this type of data. This feature supports later on in the formulation of queries and application programs across data units by combing associated operations of different modules into higher level operation and ultimately into application programs.

The data definition language enforces the definition of elementary building blocks of application programs along with the data they are referring to in module descriptions. The composition mechanism that is provided with the data definition language enables the construction of modules in a hierarchic fashion. This allows the definition of arbitrary data composition in each schema and the construction of application programs out of elementary queries associated with these data within a module.

The schema definition and application program generation function may be depicted as follows:

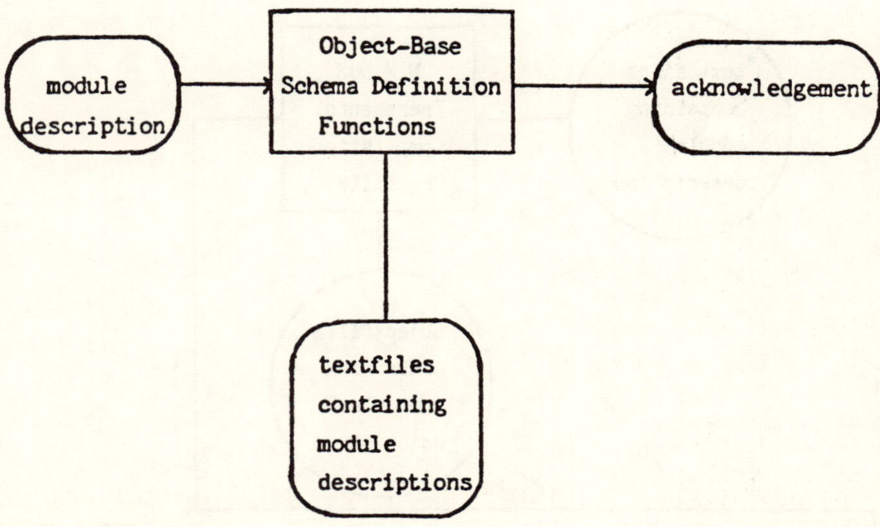

Incremental compilation

After the definition of a data base schema in terms of modules and after the hierarchical constructions of modules the resulting module hierarchy may be compiled with the compilation function provided in Object-Base. After its compilation the schema consists of executable modules, i.e. the compiled schema contains now object versions of the data descriptions and object versions of the application programs. The execution of the compiled application programs may be initiated through the query processing function provided by the system (see section on query processing function).

The compilation function consists in fact of two subfunctions: one that compiles individual module descriptions and a second one – the binding function – that links a newly compiled module into the already existing module hierarchy at its proper place.

High level query execution

The system provides high level query facilities to support a variety of different users with their own external schema and their own collection of application programs associated with their external schema.

Queries are not formulated by the end-user but rather by a database expert at schema definition time (see section on schema definition). They may be formulated at a level of abstraction most suitable for a specific application. Each external schema represents, therefore, an application tailored query interface.

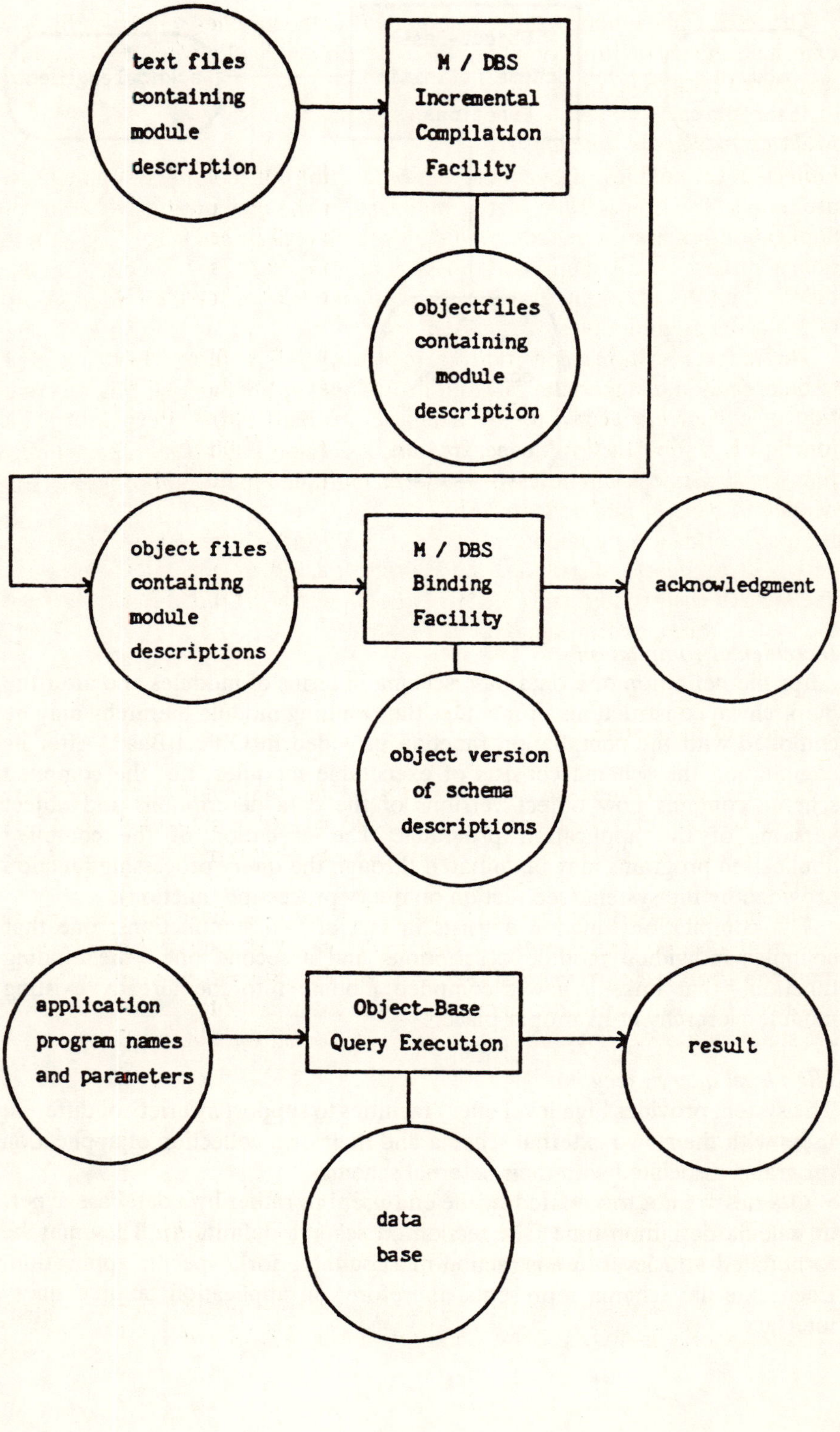

text files containing module description

M / DBS Incremental Compilation Facility

objectfiles containing module description

object files containing module descriptions

M / DBS Binding Facility

acknowledgment

object version of schema descriptions

application program names and parameters

Object-Base Query Execution

result

data base

The user only initiates the execution of a precompiled query at his/her external schema by supplying the names of the applications programs and its associated parameters.

Multi-user database functions
Object-Base enables the concurrent execution of compiled application programs. The consistency of the database in the concurrent execution of application programs over common data will be guaranteed since each simple query program and composed application program (see section on the Application Programming Interface) is designed to guarantee the consistency of the associated modules' data type.

This in fact results in a new transaction concept and requires synchronisation to be organised in a modular fashion. Provisions for the modular synchronisation of concurrent access to the database are built into Object-Base. This function is a sub-function to the previously defined High-level-Query-Function whose services will be used whenever multiple application programs are needed to be executed concurrently.

Since workstation computers are frequently interconnected in local area networks to enable distributed computing, data management services are needed to be distributed as well. Provisions are made in Object-Base to extend the system into a distributed database system.

5. Status of the system

A first version of Object-Base has been developed at the University of Bremen [Weber 84]. The development team consisted mainly of students working on the project in partial fulfilment of the curriculum requirements for a Diploma degree.

The project lasted for two years and led to the specification of the entire system and to the implementation of a rudimentary version of it (running on the VMS operating system on a VAX 11/750). The completion of the system is planned. The purpose of the first version is to act as prototype and so to verify the statements that, hopefully, have been made plausible in this paper.

6. Summary

Starting with the concept of abstract data types a new kind of database system, Object-Base, is developed. The concept turns out to be very fruitful as it leads to many advantageous features of Object-Base.

- The system is adjustable to very different application characteristics. Thus it also supports in-situ database facilities on workstation computers.

- Complex objects are composed hierarchically out of simplex objects. High levels of semantic data models can be provided.
- For truly casual users a push-button interface can be integrated into the application module hierarchy constituting the conceptual schema with the operations allowed on the data.
- The schema description can be modified easily by addition, deletion and replacement of components (modules).

Because of the object-orientation of Object-Base we think the system to be relevant for the development of future database systems.

References

ANSI 75 'Interim Report ANSI/X3/SPARC, Groups in Data Base Management Systems', FDT-ACM SIGMOD Bulletin H.2

Liskov 75 LISKOV, B., ZILLES, S.: 'Specification Techniques for Data Abstractions', SIGPLAN Notices 10, 6, June 1975, pp. 75–83

Weber 83 WEBER, H.: 'Object-Oriented DDBS-Design', ICOD 1983, Cambridge

Weber 84 WEBER, H., SEEHUSEN S.: 'Entwicklung eines modularen Datenbankverwaltungs-systems (M/DBVS/)', Abschlußbericht, Projektgruppe DBVS, Technical Report 8/84, University of Bremen, Informatik (in German)

Wirth 82 WIRTH, N.: 'Programming in Modula-2', Springer Verlag, Berlin, 1982

Structuring mechanisms in distributed systems

Radu Popescu-Zeletin
Hahn-Meitner-Institut for Nuclear Research, Glienicker Str. 100, 1000 Berlin 39, Germany

Abstract

The paper reviews some of the basic structuring mechanisms in distributed systems from the perspective of an wide-accepted reference model (ISO/OSI). The acceptance of the model and its related standards by international bodies (ISO, IEEE, CCITT, IEC etc.) *and* the computer manufacturers will have an important role in the development of distributed systems based on available products. The paper outlines the general framework and the requirements of distributed systems and focus on the practical and conceptual problems in using the ISO/OSI in the design of distributed systems.

1. Introduction

Not very long ago, the interface between the user's device and the modem was generally regarded as the line dividing data communication and data processing. The past decade the network has crept through that connector and has infiltrated in terminals, main frames and frontends.

The user needs and a growing market of data communication have precipitated this invasion by calling for increased connectivity, higher reliability, lower costs in networking, support for interconnecting heterogeneous devices and development of systems and applications dealing with communication [1].

The result of this invasion is a large variety of products, architectures and systems for distributed applications. Recognising that the user and the manufacturer community have a difficult task to get unscathed through the jungle of concepts and products, different standardisation bodies have developed reference models for coherent development and integration of concepts and products in the field of data communication.

The paper outlines the major structuring techniques and analyses the present output with respect to distributed systems requirements.

Chapter 2 gives a short overview on the rationales of the ISO-OSI model and the used tools for hierarchical structuring.

Chapter 3 outlines the main structuring mechanisms and their rationales.

Chapter 4 describes from a personal point of view the state of the art and what is needed and missing with respect to distributed systems.

2. The reference model for distributed systems

One of the major problem of distributed systems is their inherent complexity. This complexity is motivated by the combination of the traditional fields in informatics: data processing and data transmission.

The close assembling of the two fields to form distributed systems despite advantages like: increased availability, parallelism, increased reliability and performance, requires the reconsideration of the problems and solutions in both fields. The aim is a distributed system architecture where data-transmission and data processing are melting in one.

The complexity of distributed systems is motivated by:

- If the traditional data processing systems are designed for a certain configuration, the use of DP-systems combined with data networks permits a large variety of configurations to cover a large variety of applications.
- If the traditional telecommunications networks offer highly specialised services (teletex, videotex, terminal access etc.), the distributed processing systems are required to perform a variety of functions modelling a growing field of applications and requirements.
- If the traditional data-processing is based mainly on sequentiality, the natural starting point for distributed systems is parallelism. Parallelism introduces a new way of thinking in problem formulation and solving.
- Expected advantages of distributed systems like: higher availability, reliability and performance are paid internally by complex algorithms which are characteristic for distributed control of resources [2].

The well-known technique for solving complex problems is the decomposition of the initial problem in a model of less complex problems also known as the architecture of the system.

At international level different standardisation bodies have tried to develop systems architectures with reference character for complex systems. One example is the ANSI/SPARC model for database systems.

For distributed processing systems the international efforts have merged in the well-accepted ISO-Open System Interconnection Reference Model (ISO/OSI).

It is probably interesting to analyse the term: Open System Interconnection Reference Model, because it hides the aim of the model. The scope is to interconnect systems which are open in their architecture to co-exist and co-operate with other similar systems. The key issue is a common architecture which is flexible enough to allow changes without throwing away the whole system.

The method consists in describing a model of a network of Open Systems (necessary for OSI standard designers) as a network of models of Open Systems (necessary for implementing an individual Open System).

The second characteristics is the reference character of the developed architecture, which allows to develop components and products with reference to an accepted reference model.

That means that the reference architecture must provide a precise frame-work in which the internal functionality of Distributed Systems is clearly decomposed in independent components which encapsulate functions without influencing each other.

The precise formulation of the model address the external visibility of each component (layer) and the relationship between the components of the model.

This technique allows products development for the different components in a disciplined way.

The modelling technique of encapsulating related functions in modules and providing autonomous external visibility is close to the well-known concepts of abstract data types. The resulting architecture is a layered one which allows as well the discrete evolution of hardware and software in the different layers, the introduction of new systems without making previous implementations obsolete.

The reference character of the OSI model has already and will have a dominant role in future distributed systems since it imposes a rigorous discipline by providing a logical frame-work for a complex domain and for the development of products and systems to achieve compatibility.

The adapted layering technique is characterised by two major concepts (Fig. 79).

- *The layer service:* which is the set of capabilities offered at the boundary of a layer to a user in the next higher layer. Note that the service is an abstraction by which the capabilities offered by a layer (in using all lower layers) are specified and that the service definition is independent of any particular implementation.

- *The protocol* which defines the rules of interactions between the entities which are situated in the same layer but pertain to different systems.

  It is obvious that by the definition of the service offered by a layer and the definition of the service offered by the lower layer the functionality of a layer is completely specified. Note that different protocols may carry out this functionality.

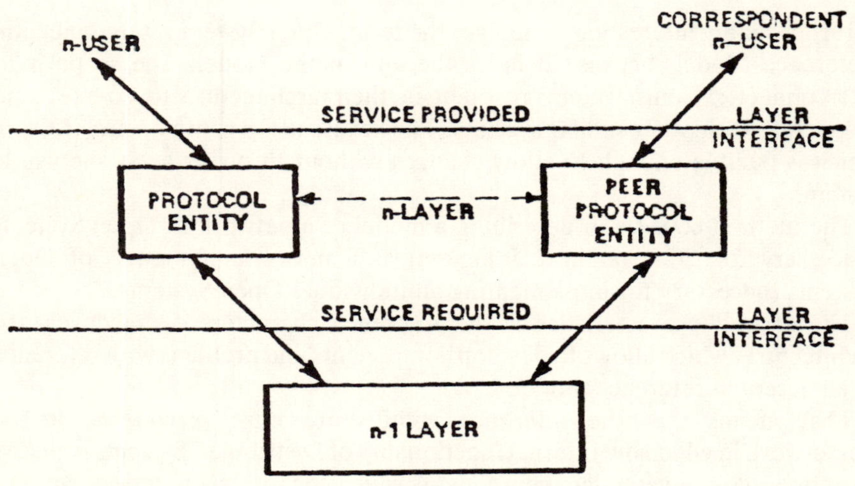

Fig. 79 *Services and protocols in ISO/OSI*

The result of the conceptual analysis of distributed systems functionality is a seven-layer architecture. One of the key issues of the model is the separation between data-transmission and data-processing domains at the boundary of layer 4.

The first four layers provide a uniform communication kernel and deal with functions necessary to provide and support a diversity of topologies, error recognition and recovery mechanisms, efficient transmission and costs optimisations.

A firm end-to-end basis for inter-process communication in which all the above problems of data transmission are hidden for the user is an essential feature for the development of distributed systems.

Note that in this chapter we do not address the different products and standards in the different layers of the ISO-OSI their advantages and their weak points but the rationales for developing a reference model and the model itself.

The upper three layers are probably the most interesting since they address a domain which was influenced mainly by the traditional operating systems and basic constructs of high-level languages.

OSI brings new influences in these functions focusing on communication and dialogue aspects which have been poorly treated in the past.

Fig. 80 focuses on an informal comparison among functions offered by operating systems and programming languages on one hand and functions offered by the Session and Presentation layer as described now in the standards [2].

TRADITIONAL DATA-PROCESSING		O. S. I.	
Operating Systems functions	Prog. Languages functions	Session Layer functions	Pres. Layer functions
RUN		Estab. Sess.	
Enqueue/Dequeue		Send/Receive	
Post / Wait		S/R Expedited	
Semaphore		Token	
	Co-routines	TWA dialogue	
	Procedure Call	(Part of TWA)	
	Data Types		Syntax
	Declarations		Syntax Negotiat.
	End	Release Sess.	
Check Points		Sync.	
Restart		Resync.	

Fig. 80 *Analogy OS/Prog. languages and ISO/OSI (from H. Zimmermann 'On Protocol Engineering')*

3. Structuring criteria in distributed systems

Basically there are some general criteria for structuring distributed systems. These criteria are:

- Space structuring
- Time structuring
- Data structuring.

The structuring process in the distributed system design is depicted in Fig. 81.

3.1 Structuring in space

One of the important structuring criteria of a distributed system is its structuring in space. By structuring in space we *do not* mean a mapping of the system in a certain topology but rather a logical structuring in autonomous components. It is important to underline the independence of the structure from the topology since this is necessary for the open characteristics of the system.

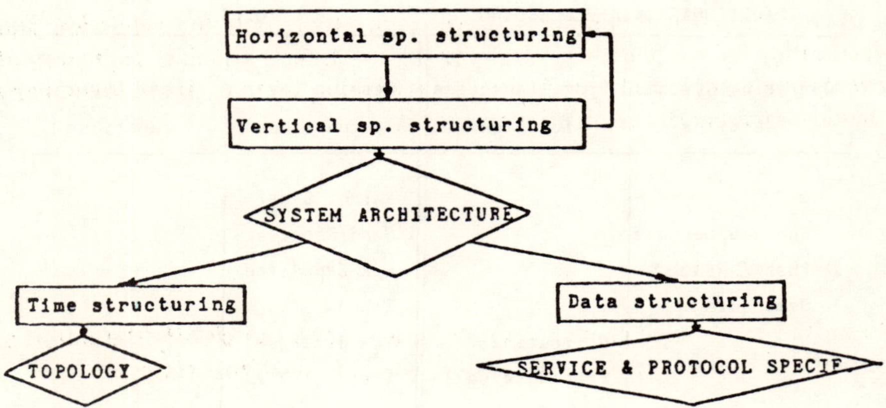

Fig. 81 *Structuring mechanisms in distributed systems*

It is also important to perform first a horizontal structuring in space in components at the same level encapsulating functions from a top-down design approach. The vertical space structuring will then define the necessary functions in the layered architecture. Note that OSI provide till now only vertical structuring and only very few provisions for the horizontal one (MHS, JTM).

It is clear that the actual developed standards in OSI and the related products model co-operation between autonomous systems and provide a framework to bridge mainly distances and heterogenity

The horizontal space structuring is necessary for a distributed application where components are tightly communicating to provide one single application. In this class of distributed systems are the fault-tolerant systems, resource sharing systems, real-time applications and systems designed for high-reliability and performance. The space structuring must provide horizontally in each layer explicit address-spaces, which are governed by explicit protocols.

3.2 Time structuring

In order to be able to cope with the parallelism in distributed systems an important aspect is the structuring in time of different activities. This allows not only to exploit the distribution by allowing independent, parallel processing but also to define the best software and hardware topology.

Tightly coupled with time structuring are aspects like synchronisation of parallel processing, operation, atomicity and data consistency.

Again, the related ISO-OSI standards provide at this time only mechanisms for co-operation of point-to-point autonomous applications and the user of the model has to deal with more sophisticated communication aspects outside the model.

A very important point is the requirement of tools for validation and verification of the correctness of the protocols and possible sequences of events in a complex environment. The availability of tools has to accompany the development of the distributed systems.

3.3 Data structuring

From the experience gained in the last years, one can classify the different structuring approaches of a distributed system in two gross classes:

- Program-oriented
- Data-oriented.

Comparing the two approaches it seems that the data-oriented one has a lot of advantages. By a precise definition of the data structures in each level of abstraction we gain clarity in the distribution.

The OSI-model provides a clear separation of the protocol elements pertaining to a certain level and transparent data from/for the level above. This supports the independence of the layers in the model. The use of abstract data types concept to model the different levels and components in a distributed system is appealing in this context.

Tightly coupled with data structuring principles are error-detection and recovery mechanisms. Walker made already 1977 the observation:

> 'For a large majority of applications, it is much more cost effective to expect failure to occur and to recover from them than to aim for a totally fault-free system.'

The encapsulation of errors in structural components allows the development of error recovery mechanisms and thus provides a higher reliability in the systems. The developed products and standards in ISO/OSI follow already these principles.

A good illustration is the design of the transport protocol classes 1 and 3 where the life time of a transport connection may span more than one network connection using recovery mechanisms from network failures.

4. State of the art

The development of standards and the products within the OSI model have been influenced by the immediate market. They mirror a certain class of applications in the field of office automation and telematic services.

The field of applications which is now covered by the standards have the following characteristics:

- The systems involved are autonomous (horizontal space structuring – same functionality in each system).
- The applications in autonomous systems co-operate in point-to-point communication regime.
- The different standards for each level have been developed to bridge distances and heterogeneity.

Fig. 82 depicts the standards for each level.

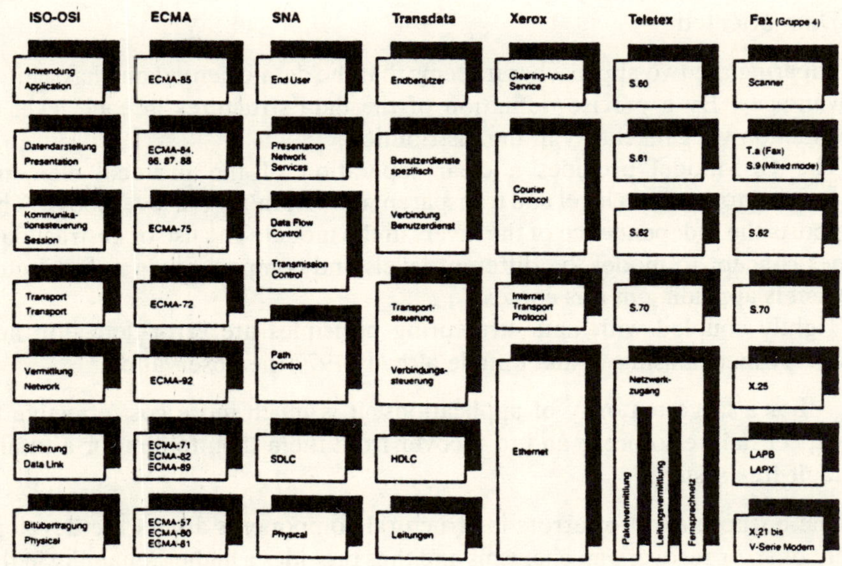

Fig. 82 *Standards supported by different organisations (from DATACOM 2/84)*

4.1 Open systems

The term open-systems hides the wish that each system will conform to the reference model and will be able to communicate with all other systems. Since nobody can foresee all the future applications to be supported by these systems the wish is quite utopic. It is interesting to observe which are the mechanisms to support this wish.

The most important one is the provision of negotiating at service, and consequently at protocol level the required quality of service. Note that all standards at the different levels provide this capability.

The quality of service negotiation is specific for each level since each level is designed to perform a certain functionality and is represented by entities in a certain address space.

Note also that the negotiation takes place between at least three partners: the two users of a service and the layer below and in that negotiation all three partners may intervene. The result is an agreed communication quality during the data transfer phase. The quality of service is not the only subject for negotiation. Since powerful systems may communicate with weak systems the negotiation includes the set of primitives, subsets and functional units supported by the two open systems in order to establish the common denominator in communication.

That means that a system in order to be an open system has to fulfil at least some minimal requirements and that all systems have to be designed in such a way that they can degrade their functionality if necessary down to a well-defined minimum.

A flexible negotiation and the ability to degrade to a level of functionality imposed by the peer system is a new quality in software and hardware product design. The negotiation rules of upgrading/degrading a system are stated as general conformance rules for each level.

4.2 Protocol engineering

The above described techniques have been successfully applied in the design of standards and their associated products in the ISO-OSI environment.

A new discipline has evolved which from literature [4,2] is known as 'protocol engineering'.

The term *new* is as always relative but it is important to note that it is the only field in which:

- a general model was developed and finalised before specific standards and related products have been produced
- the development of the model and its related standards and products have been accompanied by the development of formal description technique, certification and validation tools.

The domain of 'protocol engineering' is depicted in Fig. 83.

The development of standards for the distributed system follow in their implementation some very precise steps, which mirror on one hand the modularisation and hierarchical structuring and on the other hand the involvement of different protocol engineering techniques and tools.

The wide acceptance of the ISO-OSI model and concepts by international organisations like CCITT, IEEE, ECMA, IEC and the different product suppliers is an important hint for the relevance of the model in the future.

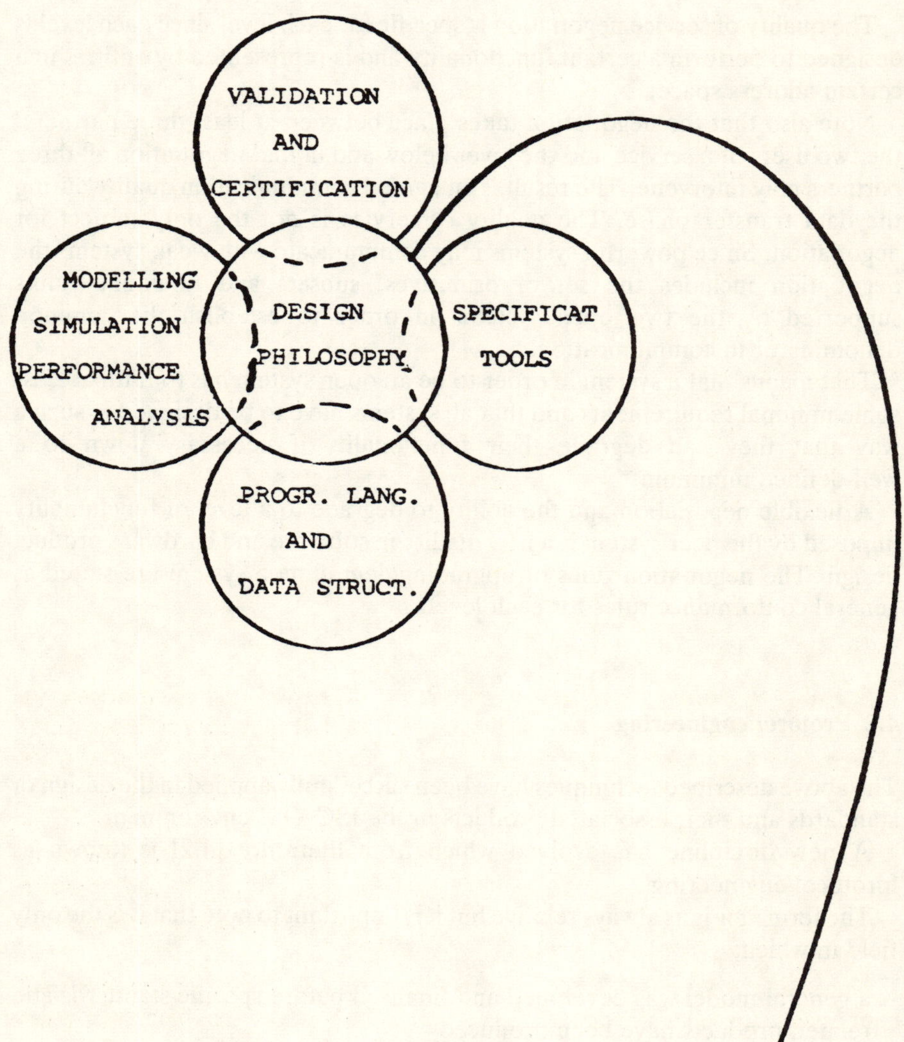

Fig. 83 *The protocol-engineering domain*

4.4 Some conceptual and practical problems

There are a number of general issues which are not supported or not clear in the ISO-OSI reference model specification and its related standards. These issues appear when the network designer intends to develop implementation specifications which are required to conform with the standards and at the same time to offer practical solutions for the network design [5].

1. *Layer independence*

As already mentioned the adopted layering technique is characterised by two major concepts:

- the service definition describing the external visibility of a certain layer and
- the protocol definition describing the internal functioning of the layer.

Although these architectural concepts aim to layer independence the services defined in the related standards for each layer do not provide completely this goal. This is mainly due to the fact that the services are defined as a three party communication. The three communicating entities are the two users of the service and the layer below as service provider.

The implication of this fact is that the specifications of the services provided by each layer define the capabilities offered by the layer *and* the behaviour of the two users of the service in the layer *above*.

A neutral service definition without the involvement of the behaviour of the entities in the layer above is probably the only way to achieve the aimed layer independence.

2. *Multicast communication*

The ISO-OSI model provides provisions for point-to-point communication only. A large class of applications requires multi-cast communication regimes (distributed data-bases, mailing and tele-conferencing systems etc.). At the present time the user of the model has to solve the communication aspects for multi-cast applications outside the model. Multicast communication introduces not only new addressing capabilities but also synchronisation mechanisms to preserve the consistency of data, and the atomicity of operations which are poorly treated or completely missing in the actual proposals. These aspects are essential for a large variety of applications.

Since one of the aims of the model is to relieve the users of the model of communication-oriented aspects, for this class of applications the model failed its aim.

3. *Layer entities*

Another practical and architectural problem is the fact that in the OSI Reference Model the existence of Entities in the next higher layer is assumed at connection set-up time (e.g. Session Entities for the Transport Layer in the establishment phase). This assumption is not true for most implementations, if the implementator cares for implementation efficiency and so does not implement a multitude of dummy processes which have to wait to be activated. That means that before a CONNECT indication may occur the creation of an entity in the Session Layer has to be performed.

It is not clear from the service specification which entity in which layer has to enter the termination phase if the entity cannot be created or when a deadlock situation occurs.

4. *Dynamic change of the quality of service*

The present documents specify the negotiation of the service quality only at connection set-up time. There are many cases where a dynamic change of the quality of service is required. In the existing documents a quality of service may be changed only by involving the termination of the connection and then by re-establishing a new connection with the new service quality. This scheme is too rigid and too costly.

A practical problem in the network design is also the criteria choice on how different quality of services at different layers can be mapped.

5. *Quality of service parameters*

Some of the quality of service specified in the ISO draft proposals as parameters of the service primitives are difficult to interpret if not impossible to support. For example, the meaning of the negotiation of the connection establishment failure probability at the connection set-up time or of the DISCONNECT failure probability is not clear for the user of the service. Other parameters can be supported only in conjunction with a powerful network management protocol specific to the layer offering the service.

The requirement to guarantee a certain quality of service is mandatory for a large class of applications (e.g. real-time). The present standards and products cannot ensure a quality of service during the life-time of a connection. The layer providing the required quality can only 'do its best' without guarantee.

6. *Performance*

Very often the network designer is faced with the question 'Does the implementation of the seven layers affect the network performance and if yes how much?'

The answer of the question is not easy, because the pilot implementations build a *network operating system* on top of the existing operating systems in each host. The scope of the standards in the OSI-environment is that the computer manufacturer has to integrate these featues in their operating systems for communication purposes.

The integration process has already begun and many manufacturers provide already computers and workstations with OS architectures based on the OSI-standards.

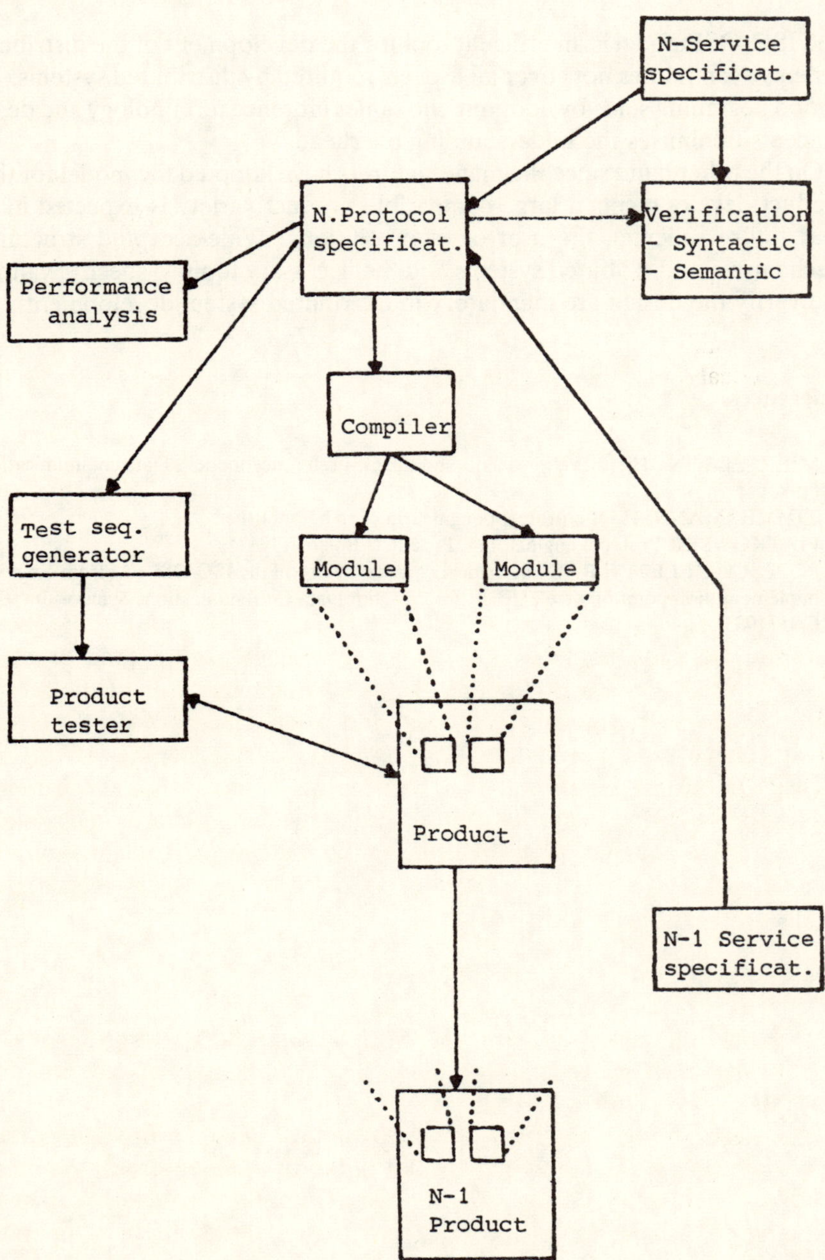

Fig. 84 *Hierarchical development of distributed system*

Conclusions

The ISO-OSI model is an efficient tool for the development of the distributed systems even if does not cover all aspects required by distributed systems. It is a good beginning since by adopting the same reference terminology and design concepts minimises the understanding overhead.

On the other hand since the manufacturers have adopted the model for their product developments a large compatible product variety is expected in the near future relieving the user of own solutions. Wide-accepted structuring mechanisms of distributed systems and the necessary tools to specify, validate and verify the design are mandatory in distributed system development.

References

1 MIER, EDWIN. 'High-level protocols and the OSI reference model', Data communications. (1982)
2 ZIMMERMANN, H. 'On protocol engineering', to be published
3 PIATKOWSKI. Protocol engineering, Proc ICC Boston (1982)
4 POPESCU-ZELETIN, R. 'Some critical considerations on the ISO/OSI RM from a network implementation point of view', IEEE Proc. Eighth Data Communications Symposium, Cape Cod (1983)

Fault tolerant systems in military applications

M. R. Moulding

Royal Military College of Science, Shrivenham, UK

Abstract

This paper introduces some basic principles and terminology associated with the design of highly reliable computer systems, and describes two experimental, fault tolerant computer systems which have been recently developed for military applications. The first system, called ADNET, demonstrates how a dynamically reconfigurable, local area network can be constructed so that various forms of hardware failure can be tolerated. The second system illustrates how software fault tolerance techniques can be used in a real-time application in order to cope with software design faults and, thereby, improve the software reliability of the system.

1. Introduction

Computer systems are often used to perform critical functions where their assured reliability is of paramount importance. Such applications have traditionally been associated with military and aerospace agencies which have funded much of the research into the development of highly reliable computer systems. The notion of incorporating protective redundancy into a system as a means of tolerating operational faults has emerged from this work to become a well-established technique for achieving high reliability. Such redundancy normally involves the inclusion of extra hardware units, additional software and spare processing time, and represents the price paid for increased reliability.

During the last decade, the reducing cost of hardware has led to the widespread use of computers in general industrial and commercial systems, and a growing number of these applications have stringent reliability requirements which demand the adoption of fault tolerance techniques.

Consequently, there is an increasing need for industrial and commercial system designers to appreciate the architectural features of modern fault tolerant computers and it is the purpose of this paper to provide an introduction to this technology. In order to eliminate a possible source of misunderstanding, we shall first start with a brief discussion of some basic reliability concepts and terms.

2. Terminology

The field of computer system reliability brings together a number of disparate professional groups (e.g. hardware designers, control engineers, software engineers) which each have their own set of terms to describe essentially the same reliability concepts. This can lead to a great deal of confusion and fruitless debate which prevents a meaningful discussion of the underlying concepts. The reconciliation of this problem is beyond the scope of this paper but, in order to discuss some basic reliability issues, a consistent set of terms is required. We shall base our terminology on that provided by reference 1 in the knowledge that not all aspects of it are universally accepted.

Generally, the 'reliability' of a system is characterised by a mathematical function R(t) which expresses the probability that a system will not fail during

TERMINOLOGY (1)

RELIABILITY	IS CHARACTERISED BY A FUNCTION R(t) WHICH EXPRESSES THE PROBABILITY THAT A SYSTEM WILL NOT FAIL THROUGHOUT A PERIOD OF DURATION t
MTBF	MEAN TIME BETWEEN FAILURES
FAILURE	OF A SYSTEM OCCURS WHEN THE BEHAVIOUR OF THE SYSTEM FIRST DEVIATES FROM THAT REQUIRED BY ITS SPECIFICATION
EXACT SPECIFICATION	IS REQUIRED WHICH MUST BE: CONSISTENT COMPLETE AUTHORITATIVE USABLE AS A TEST FOR FAILURE
ERROR	A DEFECTIVE VALUE IN THE STATE OF A *SYSTEM*
FAULT	A DEFECTIVE VALUE IN THE INTERNAL STATE OF A *COMPONENT* OR IN THE STATE OF A *DESIGN*.

Fig. 85

a specified time interval. For operational systems, this function cannot be known but by using reliability modelling techniques [2, 3] its form may be predicted and the values of its parameters estimated. If such a modelling exercise can be carried out successfully, then it is possible to estimate the probability of a system failing during a defined operational period. This, however, is not a commonly used reliability metric since it can only be accessed via an appropriate model. Instead, mean time between failure (MTBF) is often used since it can be measured directly for an operational system by recording system failures, or obtained from a suitable model which itself may often require recorded failure data to refine its prediction.

Clearly, any attempt to assess objectively the reliability of a system requires a precise definition of what constitutes system 'failure'. For this we require a specification detailing the precise functionality of the system. A 'failure' can then be said to occur when the behaviour of a system first deviates from that defined by its specification. Such an innocuous definition belies the immense difficulty of producing a specification adequate for this purpose. The specification must be internally consistent; it must be complete in the sense that all possible operating conditions and responses are catered for; it must be authoritative so that judgements derived from it are unquestionable; and it must be expressed in a way which allows it to be used always as a test for system failure. A specification possessing these properties is termed an 'exact specification' and since such specifications rarely exist we must accept that the definition of failure, and hence the assessment of reliability, will usually contain a degree of subjectivity.

Possibly two of the most frequently used reliability terms are 'error' and 'fault' and in order to allocate precise definitions for them, it is necessary first to construct a simple model of a computer system. Generally, we can consider a computer system to contain a hardware system and a software system which

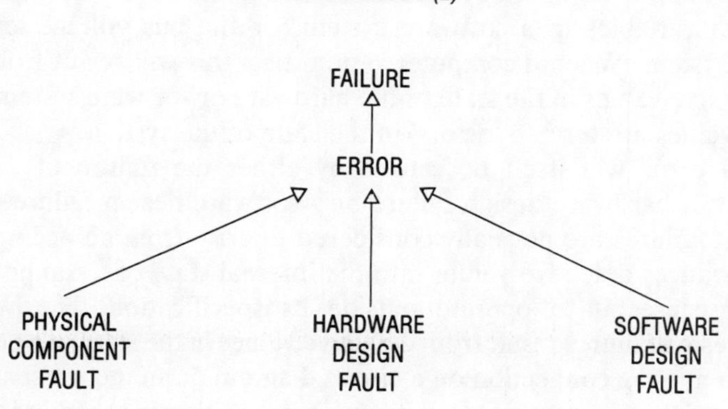

Fig. 86

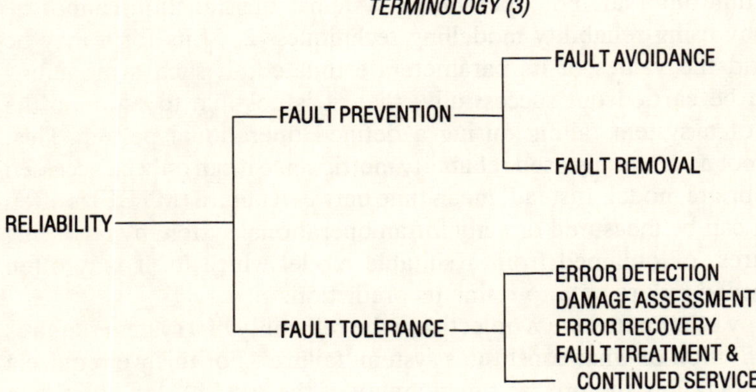

Fig. 87

combine to perform the external functions of the total system. Each of these logically separate systems will contain a number of 'components' which interact under the control of a 'design'. Components may themselves be considered as systems in their own right and contain sub-components interacting under a design to perform the overall functions of the component. This hierarchical decomposition will continue until 'atomic' components, which are considered to have no internal design, are identified. In the case of a hardware system, decomposition will result in a series of designs and a set of physical (atomic) components; a software system will decompose purely into a set of designs, since software has no physical properties. During operation, a system will adopt a number of distinct internal states; for example, the values of program variables in a software system, or the bus voltage levels in a hardware system. When a computer system fails this will result from one or more defective values in the state of the hardware or software systems. These defective values are termed 'errors' in the state of the system.

Such an error will itself be caused by either the failure of a physical component, a hardware design failure or a software design failure. Physical component failures are normally considered to arise from an ageing process which introduces defective values into the internal state of a component and eventually causes it to operate outside its specification. Hardware and software design failures result from defective values in the state of a design; for example, a missing connection on a circuit diagram or an incorrect statement in the source text of a program. A defective value, either in the internal state of a component or in the state of a design, which causes an error in the system state, will be viewed as a 'fault' in the system.

From the discussion above we can derive a relatively simple model for computer system failure. During normal operation, the activity of the system may be such that a physical component fault, or a residual hardware or software design fault, is encountered and damages the system state by generating one or more errors. Again, depending on the precise activity of the system, an error may cause the external behaviour of the system to deviate from its specification and so cause a system failure. There are two complementary techniques which can be used to reduce the possibility of system failure and so provide high reliability:

(i) Fault prevention. The aim of this technique is to try to prevent faults from existing in an operational system. There are two separate approaches:

   (a) Fault avoidance. The objective here is to try to avoid the introduction of faults into the system. For example, design faults can be reduced by the adoption of a good design method; physical component faults can be reduced by the use of top quality components.

   (b) Fault removal. This method assumes that faults will have been introduced during the development of a system and strives to remove as many as possible, by exhaustive validation and testing, before the system is launched into service.

(ii) Fault tolerance. If prevention schemes cannot provide the required reliability for a computer system, then it will be necessary to construct the hardware and software systems in such a way that they can prevent a fault from causing system failure. Such an approach requires a combination of the following activities to be carried out by the system:

   (a) Error detection. The presence of the fault cannot be detected directly; it is the detection of an error in the system state which identifies the presence of a fault and can be used to instigate corrective action.

   (b) Damage assessment. Following the detection of an error, the extent of the damage to the system state may be estimated. Such estimates are normally based upon static damage confinement structures which exist within the system.

   (c) Error recovery. Before a system can be allowed to operate normally following the detection of an error, the system must be returned to an error-free state.

   (d) Fault treatment. If a system is allowed to continue normal service following error recovery, it is possible that the fault will recur and lead to eventual system failure. To avoid this problem it is necessary to locate the fault and remove it from the system, by some form of reconfiguration, before allowing normal service to continue.

In the following sections we shall investigate how the fault tolerance principles outlined above can be applied to hardware and software systems.

3. Hardware fault tolerance

Hardware fault tolerance schemes are invariably based on the assumption that hardware design faults will not exist and that physical component faults will be the sole cause of potential failures. The rationale for this lies in the lower complexity of hardware designs when compared with software, and the widespread use of standard, operationally proven designs for integrated circuit devices and printed circuit boards.

The ageing process of physical components can be well characterised via accelerated life testing and, consequently, the effects of a component fault are 'predictable'. This is of significant advantage when attempting to devise a fault tolerance strategy. For example, if a certain component is known to fail in a particular manner, then the resulting damage to the system state can be predicted thus facilitating error detection and recovery. Furthermore, redundant components which must be added to the system to protect against physical faults can be of the same type and design.

Redundancy in hardware systems can be categorised as either 'dynamic' or 'static' [4]. In a dynamic scheme, a faulty component will usually provide some level of assistance with error detection but will rely on its surrounding environment to carry out the other phases necessary for fault tolerant operation. The standby sparing scheme illustrated in Fig. 88 is an example of such an approach. Here a main system component (M), periodically reports to a 'watchdog timer' (W). Should the main component fail to report within a specified time interval then this will be recognised as an error by the timer which will be responsible for assessing the damage caused by the component fault, recovering the system to an error-free state and treating the fault by switching in a redundant 'standby' (S) component. The damage assessment and error recovery phases can vary in sophistication. A simple approach is to assume that the entire state is damaged and to recover it by means of a hardware reset. More elaborate strategies can lead to resets for only parts of the system, based on some a priori characterisation of the fault and/or damage confinement structures which exist in the system.

In contrast to the dynamic redundancy approach where the surrounding environment of a component plays an important role in the overall fault tolerant behaviour, the objective of static redundancy is to mask the effects of a component fault from the surrounding environment. The canonical example of static redundancy is the triple modular redundancy (TMR) unit illustrated in Fig. 88. Here, three identical components are subjected to the same inputs and the overall output is obtained by a two-out-of-three vote on the outputs of the individual components. Consequently, the TMR unit will mask the effects of any single component fault. Error detection is provided by the voting check which also locates the faulty component. Damage assessment is based on the assumption that the faulty component operates in complete isolation (termed an atomic action) and, consequently, cannot damage the system state. Error recovery simply involves ignoring the output values identified by the voting

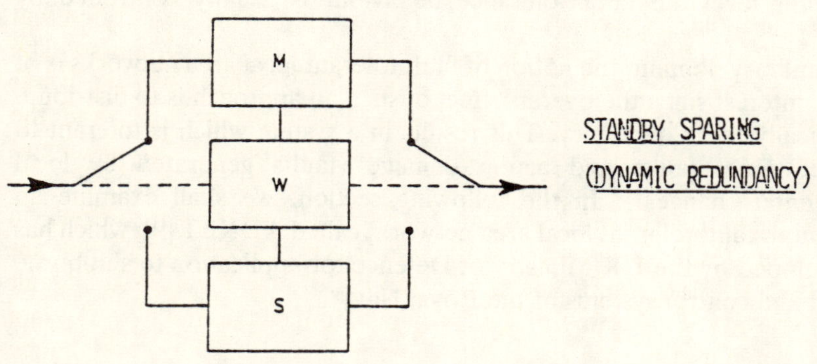

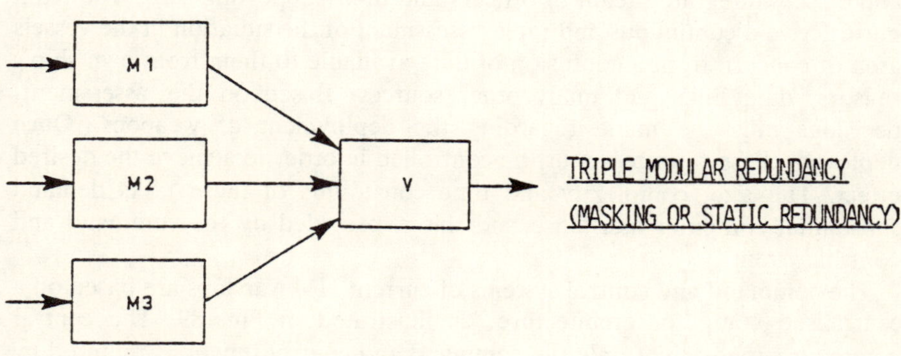

Fig. 88

check as being erroneous; fault treatment may involve ignoring future outputs from that component, depending on whether the fault is considered transient or not. Where future output from a faulty component is ignored then the TMR unit will lose its fault masking properties unless a new component is switched in to replace the faulty one.

In principle, redundancy can be applied at any level within a system. However, the higher the level at which it is applied, the larger the range of faults it protects against. The reducing costs of hardware and the increasing functionality of integrated circuits mitigates against the traditional cost penalties of this approach and a number of fault tolerant multi-processor systems have been developed (e.g. references [5, 6, 7, 8]) where redundancy is applied at the intra-computer bus level (e.g. processor and memory modules). The emergence of local area network technology also invites the application of redundancy at the inter-computer bus level (i.e. a local area network with redundant computer systems). Where redundancy is applied at the processor or computer level the fault tolerance behaviour is usually controlled by software.

In the military domain, the notion of fault tolerant local area networks is of particular interest since the overall effect of such an approach is to distribute geographically the redundancy. This results in a system which is tolerant to both operational faults and action damage (faults generated by local environmental changes!). In the following section, we shall examine an experimental, fault tolerant local area network, called ADNET [9], which has been developed by the UK Ministry of Defence for application to shipborne command and control systems of the Royal Navy.

4. The ADNET experiment

The fighting capability of present day warships is controlled by a substantial and closely integrated team of officers and their supporting staff. The team must derive a continuous and rapid assessment of the situation in the vessels area of concern from a confusion of data available to them from own ship's sensors, data links and many other sources. Based on this assessment, decisions must be made regarding the deployment of weapons. Once deployed, these resources must be controlled in order to achieve the desired effect. The size, complexity and time constraints of these tasks demand substantial computer assistance and this is provided by the command and control system.

The command and control systems of current HM warships are based on a centralised computer architecture, as illustrated in Fig. 89. The central computer supports not only the command and control functions required by the officers and their staff but also the data processing and control requirements of the ship's sensors and weapons. Such an architecture lacks enhancement capability, and is vulnerable to action damage. Consequently, a futuristic command and control system architecture has been proposed in which each weapon and sensor has computer power for local data processing, and the command and control functions are horizontally distributed across a number of computers, each with their own operator display. The proposed configuration, illustrated in Fig. 90, uses a serial data highway to provide the necessary inter-computer communications.

In order to investigate the feasibility of a horizontally distributed command and control system, the ADNET experimental model was developed at the Admiralty Surface Weapons Establishment (ASWE). A simplified schematic representation of the system is illustrated in Fig. 91. At the heart of the system is the ASWE Serial Highway [10] which is a multi-drop bus to which access is by poll and response under the control of a highway controller. Some principal features are as follows:

3 Mbit/sec signalling rate, giving a maximum useful data rate of 1.8 Mbit/sec.
An upper limit of 63 nodes over a total highway length of 300 metres.
Intelligent communication processors which interface directly into the host computer's memory and handle all the highway level data transfer protocols. The host software simply views the highway as a high integrity memory-to-memory data transfer medium.
Broadcast and point-to-point messages of variable length up to 64 bytes.
Block data transfer up to 16k bytes.
Multiple levels of error detection with subsequent recovery of lost or corrupted messages (performed automatically by the communication processor without host involvement).

The use of a passive multi-drop connection to the highway means that communications in the network will not be affected by the powering down of any computer node. The use of a highway controller does, however, represent a potential source of network failure and, to overcome this, the highway is

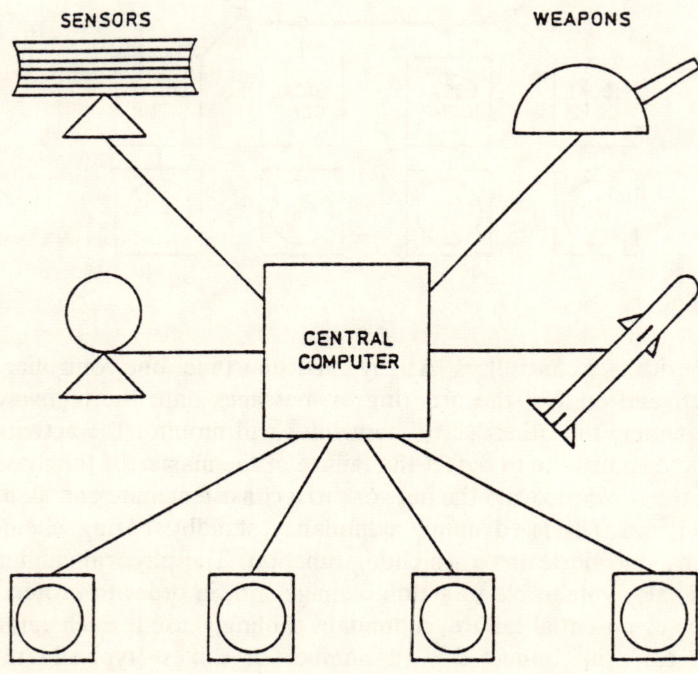

Fig. 89

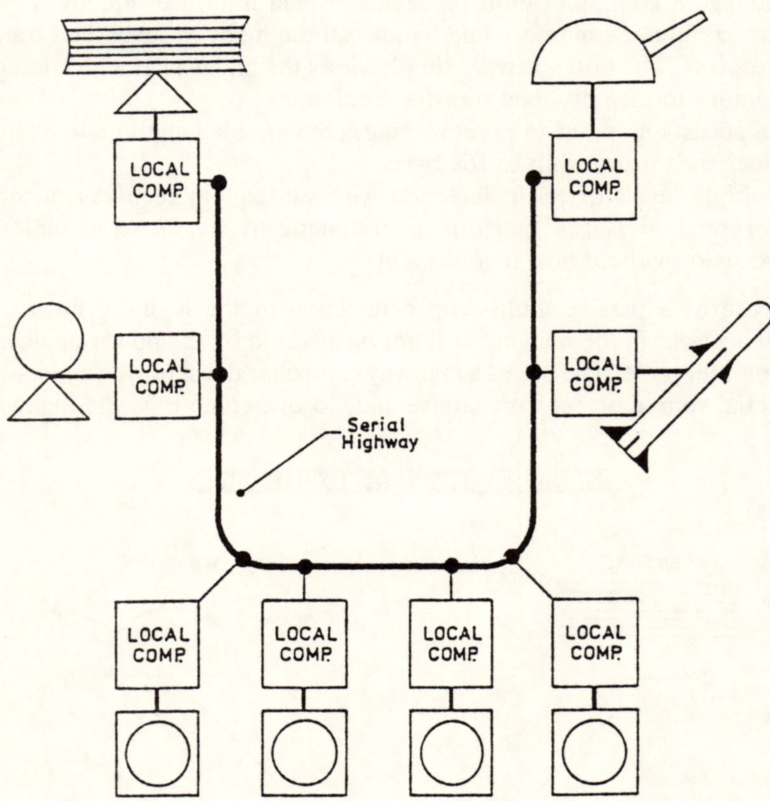

Fig. 90

equipped with two controllers. At any particular time, one controller will act as 'master' and control the ordering of messages onto the highway in the normal manner; the other 'slave' controller will monitor the activity of the highway and so be able to detect the failure of the master. If the master does fail, then the slave recovers the network to a consistent state and assumes the master function. This is a dynamic redundancy, standby sparing scheme where the standby unit performs a watchdog function. The physical cabling of the highway is also vulnerable to action damage and, in order to protect against this source of potential failure, redundant cabling is used. Each communications processor is connected to a number of cables (typically three). It transmits on all cables and selects one of that number as its input. If the reception on one particular cable has an unacceptably high error rate, then the communications processor will automatically select another cable for input.

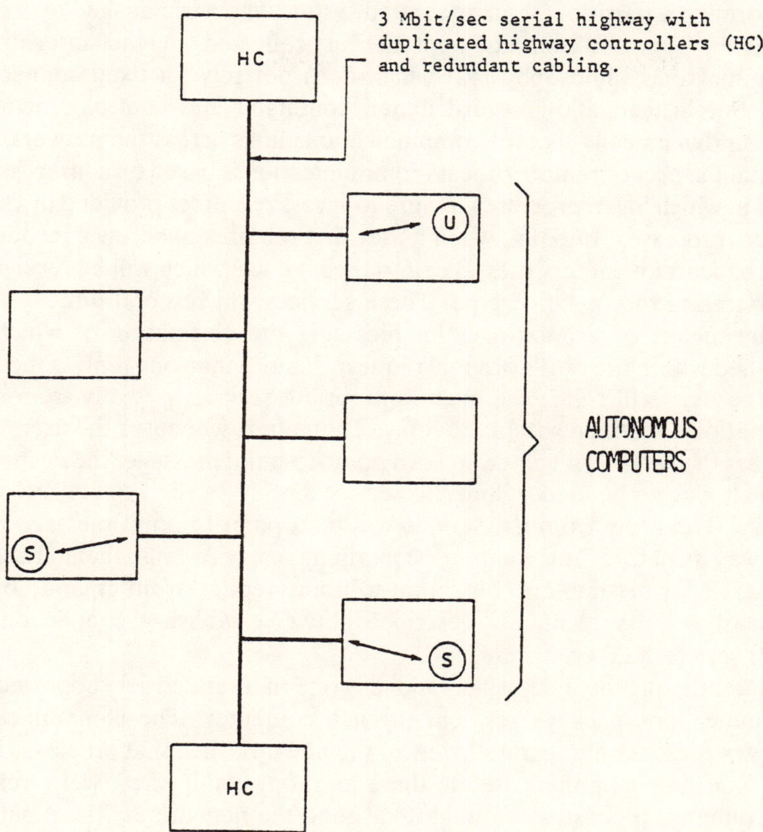

Fig. 91

Again, this is a dynamic redundancy scheme where the communication processor is responsible for implementing all phases of fault tolerance.

Connected to the highway in ADNET are a number of Ferranti Argus computers which collectively provide the command and control functions of the system via appropriate operator displays (not illustrated) in order to provide data for the command and control functions. All ADNET software is based on MASCOT [11] and programmed in CORAL.

Although the highway itself is capable of fault tolerant operation, fault tolerance at the system level will be identified by the preservation of the command and control functions in the presence of the failure of a particular computer node. This can be achieved by adding redundant computing power to the network and providing the software with the ability to reconfigure itself dynamically in the event of the failure of a particular node.

The ADNET approach to the dynamic reconfigurability problem is based on the concept that each computer should operate largely autonomously, rather than forming a partition of an integrated system which is managed by a global executive. The autonomous approach is reflected in the inter-process communications philosophy [12] which does not rely on fixed connectivity tables but instead allows a distributed command and control function to establish dynamically its own communications links across the network. One important aspect of remote process communication is based on a 'user-service' model in which 'user' processes require to access resources provided by remote 'service' processes. Initially, when a user first requires to access a resource, it will broadcast an enquiry message onto the highway which will be received by all services of the specified type. These services will respond directly to the user by means of a point-to-point message, the destination of which was contained within the user's original request. If more than one reply is received, then the user will select the most appropriate service, possibly from status information contained within the reply. Thereafter, whenever the user wishes to access the service, it will do so via a point-to-point message, the destination of which was embedded within the service's reply to the original broadcast enquiry. The reply from the service will be a point-to-point message in the same way as before. If a computer containing a service fails, then all users of that service will time-out the point-to-point replies from it and, by the broadcast enquiry technique described above, establish a connection to a similar service held elsewhere.

In fact, the protocol described above is one of a related set supported by a communications package resident in each computer. The communications packages present an integrated view of the network such that processes in the same machine communicate in the same way as if they were remote. Consequently, processes can migrate around the network and automatically re-establish their required connectivity, regardless of the particular computer they, or their communicating partners, may be resident in. This characteristic means that, providing the services that a computer supports are replicated elsewhere in the network, a particular machine may be powered down, taken out of service, powered up, reloaded with new software and then introduced back into the network, without disrupting the operation of the other machines. Such behaviour provides great flexibility in the run-time re-allocation of computer functions and significantly aids on-line maintenance.

The simple user-service model described above reveals the basic ADNET fault tolerance approach. Protective, dynamic redundancy is introduced by replicating services, which of course implies the provision of spare computing power in the network. Error detection is performed by user processes timing-out service replies. Damage assessment is based upon the physical structuring of the system since it is implicitly assumed that damage will be contained within the failed computer. Error recovery is limited to a user process disregarding any partial results obtained from the service before it failed. Fault treatment is performed by the user when it dynamically links itself to another version of the service held elsewhere.

If a service process is memoryless in the sense that its function does not depend on data retained between invocations (e.g. a mathematical function), then replicating it presents little difficulty since all service replicas will provide exactly the same function. However, if the service does retain data, then the replicas must be synchronised in some way in order that they offer the same service at all times. In ADNET such services are integrated into a specially developed, distributed database management system [13] which controls the replication, synchronisation and distribution of database partitions.

5. Software fault tolerance

In contrast to hardware fault tolerance where only physical component faults are usually considered, software fault tolerance schemes are concerned solely with design faults. This has two important ramifications:

(i) The faults and their effects will be 'unpredictable'. This increases the difficulty associated with error detection and recovery phases of fault tolerance. Backward recovery to a prior, error-free state is the most effective way of recovering from unpredictable faults.

(ii) Protective redundancy must be based on modules of independent design so as to minimise the possibility of common design faults.

The two main techniques for software fault tolerance are recovery blocks [15] and N-version programming [16]. The general syntax of a recovery block is illustrated in Fig. 92. A number of alternate modules of independent design are produced from the same specification. There will exist a primary alternate which represents the preferred design and a number of other alternates. These may be older versions of the primary (uncorrupted by enhancements), modules offering degraded functionality, or simply alternates providing the same functionality as the primary but based on different algorithms and/or produced by separate programming teams. On entry to a recovery block, a recovery point is established which allows the program to restore to this state, if required. The primary alternate is executed and an acceptance test checks for successful operation. If the acceptance test fails, then the program is recovered to the recovery point taken on entry to the recovery block, the secondary alternate is executed and the acceptance test applied again. This sequence continues until either an acceptance test is passed or all alternates have failed the acceptance test. If the acceptance test is passed, then the recovery point taken on entry is discarded and the recovery block is exited. If all alternates fail the acceptance test, then a failure exception will be raised. Since recovery blocks can be nested, then the raising of such an exception from an inner recovery block would invoke recovery in the enclosing block.

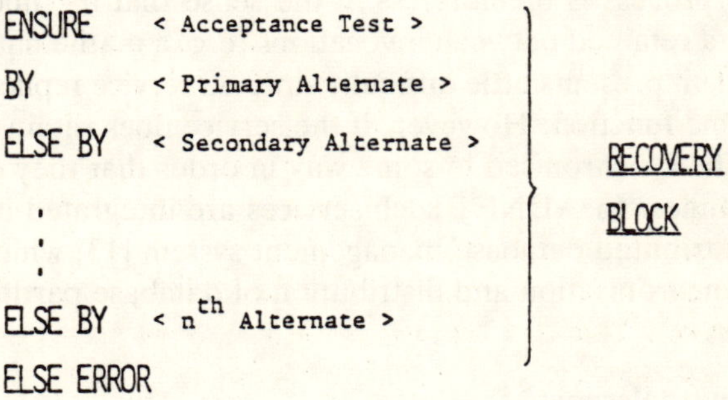

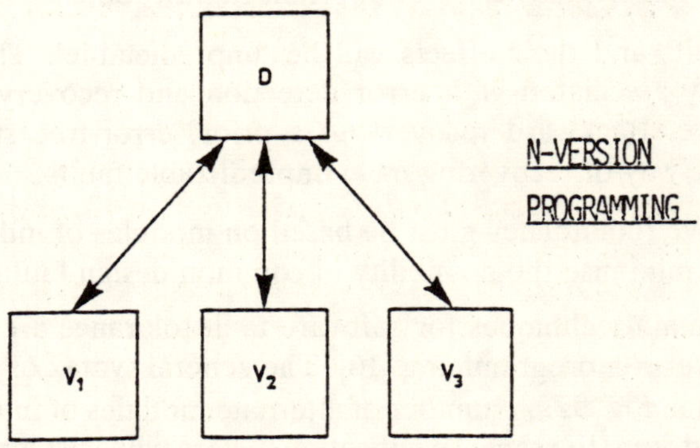

Fig. 92

Generally, an exception raised from within an alternate can be used to indicate premature failure and thus invoke the same action as for an acceptance test failure.

The recovery block approach is essentially a software analogue of the hardware standby sparing scheme described in Section 3. Redundancy is achieved by alternates of independent design; error detection is provided by the acceptance test or by an exception being raised from within an alternate. Ostensibly, damage assessment is not required because backward error recovery will eliminate all damage to the program. However, in a multiprocessing environment, backward recovery will only be applied to a single process (or at most a defined set of interacting processes – see next section) and thus practical schemes will require protection mechanisms within the machine to confine the damage to that part of the system which will be backward recovered. This constitutes implicit damage assessment. Fault treatment within a recovery block is achieved by the execution of another alternate following recovery.

In contrast to recovery blocks, the N-version programming scheme, illustrated in Fig. 92, is a software analogue of hardware triple modular redundancy. Three or more (N) independently designed versions of a module are activated by a 'driver' module (D) which supplies them with the appropriate input data. The driver then collects the individual outputs from the versions and performs a majority vote in order to determine the output from the N-version unit. Consequently, a design fault in any one module will be masked. Error detection is provided by the voting check which also locates the faulty version. Damage assessment is based on the premise that each version executes atomically (in isolation); this can be achieved physically by running each version on dedicated hardware or, logically, by running the versions on the same computer and using appropriate protection mechanisms. With atomic execution, error recovery is achieved by the driver ignoring the output values identified by the voting check as erroneous. Fault treatment can be considered as simply ignoring the results of the version identified as being faulty.

In a recovery block scheme, all alternates are always available on entry to the block, regardless of previous faults. The rationale for this is that a design fault will only be uncovered by a rare combination of processing conditions which are unlikely to recur when the recovery block is next executed. For an N-version scheme the situation is a little more complicated. Unlike the alternates of a recovery block, all versions of an N-version unit are usually executed each time the unit is invoked. Consequently, they can retain data locally between invocations. This has the advantages of increasing the design independence of the versions (alternates of a recovery block must all access the same global data structures which limits their algorithmic independence) and reducing the data which must be passed to a version upon invocation. However, if the versions do retain data, then a driver will not be able to re-use a version which has produced an erroneous output since its internal state might have become inconsistent with the others of the unit. If the fault tolerance properties of the N-version unit are not to be degraded under these circumstances, it will be necessary to provide some form of recovery of the internal state of a faulty version.

Each of the two software fault tolerance schemes described above has its own virtues. Generally, the N-version programming scheme is most appropriate to those systems which have replicated hardware for concurrent execution of versions, and for which voting checks can be easily constructed (this can be a non-trivial exercise since the versions must be of independent design and their 'correct' outputs can vary). Recovery blocks are most appropriate for systems where hardware resources are limited and voting checks are inappropriate. A full discussion of the relative merits of the two approaches can be found elsewhere [1]. The remainder of this paper will concentrate on the practical problems of using recovery blocks in real-time applications, and describe a demonstrator system, recently constructed at the University of Newcastle-upon-Tyne, to investigate the use of recovery blocks in a naval application.

6. Application of recovery blocks to real-time systems

Although recovery blocks have been available in principle since the mid 1970's, they have not been widely used in practical real-time applications. Some anticipated problems associated with their use are as follows:

(i) Run-time overhead. Acceptance tests, backward error recovery and additional alternate executions all provide a run-time overhead. Although acceptance test and alternate execution overheads are fundamental to the scheme, special hardware can be used to minimise backward error recovery times. The feasibility of this approach has been demonstrated at Newcastle University where a prototype 'recovery cache' device [14] has been developed which backward recovers the memory of a DEC PDP 11/45. The overall configuration of the device is illustrated in Fig. 94. The recovery cache is based around a DEC LSI/11 microcomputer which communicates with the PDP 11/45 host processor via a cache-host interface unit (CHIU), and can access the memory of the PDP 11/45 via a cache-memory interface unit (CMIU). The host Unibus is physically intercepted by a bus monitor unit (BMU) which is controlled from the LSI/11, and which can write data directly to the recovery cache memory via a non processor request module (NPR). Under the

USE OF RECOVERY BLOCKS

PROBLEMS ANTICIPATED:

1 RUN-TIME OVERHEAD

2 CONCURRENT PROCESSING

3 ACCEPTANCE TESTS

4 LOCATION OF RECOVERY BLOCKS

5 DEVELOPMENT COST OVERHEAD

6 MEMORY OVERHEAD

7 IMPROVEMENT IN RELIABILITY?

Fig. 93

conditions when the host processor does not require a recovery point, the BMU allows all host memory accesses to proceed unhindered. When the host instructs the cache to establish a recovery point, the LSI/11 configures the BMU to intercept all writes to memory locations which are being updated for the first time since the recovery point was established. Before these writes are allowed to proceed, the BMU reads the original value of the location and stores the location address and original value in the cache memory. It then applies the write to the host's memory. If the host instructs the cache to recover, then the LSI/11 will read the address/value pairs from its memory and restore the appropriate locations of the host's memory to their original values. In this way, the memory is returned to its state when the recovery point was established.

The operation of the cache is determined by software which runs on the LSI/11 and, in its original form, this supports four levels of nested recovery points for a single process running on the host. Initial experiments with the cache indicated that, for a typical process, the run-time overhead of monitoring the Unibus was of the order of 10%.

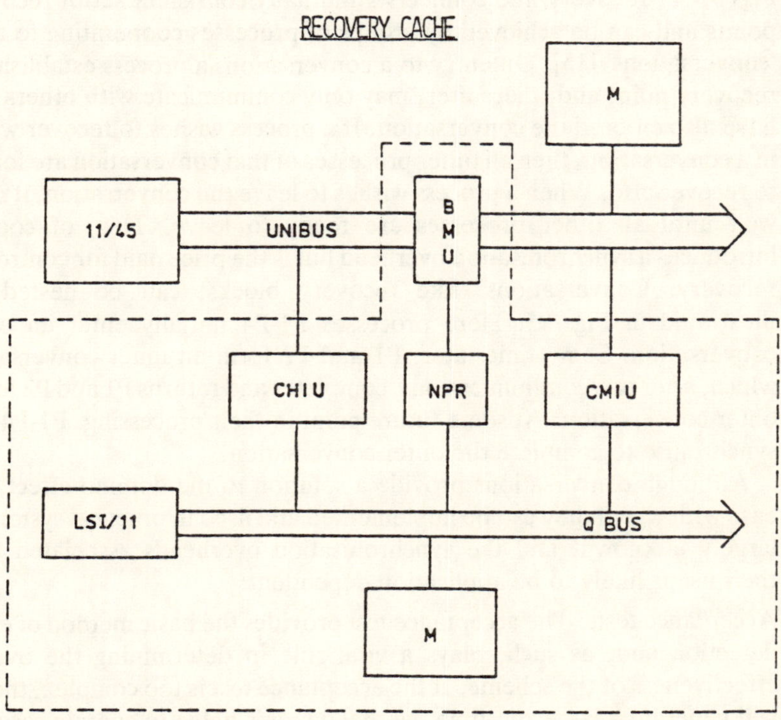

Fig. 94

(ii) Concurrent processing. When a regime of communicating processes establishes recovery points independently, then it is possible that the 'domino effect' [15] will occur. This is illustrated in the first diagram of

Fig. 95 where the horizontal lines describe the progress in time of two processes P1 and P2, the vertical lines indicate communication between processes and the open square brackets correspond to the establishment of recovery points. If, at the most advanced stage of its progress, P1 wishes to recover to its last recovery point, then this can be achieved without affecting P2. However, if process P2 wishes to recover to its last recovery point, then this will cause recovery beyond a communication with P1. In general, this communication must now be considered invalid (e.g. P2 may have passed P1 erroneous data) and so P1 must recover to its penultimate recovery point. In so doing, this invalidates further communication and causes P2 to recover to its penultimate recovery point. This sequence will continue until either a consistent pair of (possibly ancient) recovery points are found, in which case the system may proceed, or the processes will be left in an inconsistent state when all recovery points of one or both processes have been used up.

The general solution to the domino effect is to establish 'recovery lines' in the system, as illustrated by the broken lines on the second diagram of Fig. 95. A recovery line connects a mutually consistent set of recovery points and can be achieved by groups of processes cooperating to form 'conversations' [15]. On entry to a conversation, a process establishes a recovery point and, thereafter, may only communicate with others that have also entered the conversation. If a process wishes to recover whilst in a conversation, then all other processes of that conversation are forced to recover also. When a process wishes to leave the conversation, it must wait until all other processes are ready to leave. This, of course, introduces a synchronisation overhead but is the price paid for controlled recovery. Conversations, like recovery blocks, can be nested, as illustrated in Fig. 95. Here processes P1-P4 initially enter an outer conversation. Some time later, P1 and P2 form an inner conversation which, after two communications, completes and returns P1 and P2 to the outer conversation. At some future point in their processing, P1-P4 will synchronise to complete the outer conversation.

Although conversations provide a solution to the domino effect, the ease with which they can be implemented and used in practical systems is largely unknown, and the synchronisation overheads associated with their use is likely to be application dependent.

(iii) Acceptance tests. The acceptance test provides the basic method of error detection and, as such, plays a vital role in determining the overall effectiveness of the scheme. If the acceptance test is too complex, then it will generate a large run-time overhead and is liable to contain residual design faults. In contrast, a simple acceptance test may not provide an adequate method of checking the acceptability of an alternate's operation. Importantly, there is no wealth of documented practical experience upon which a designer of acceptance tests can draw.

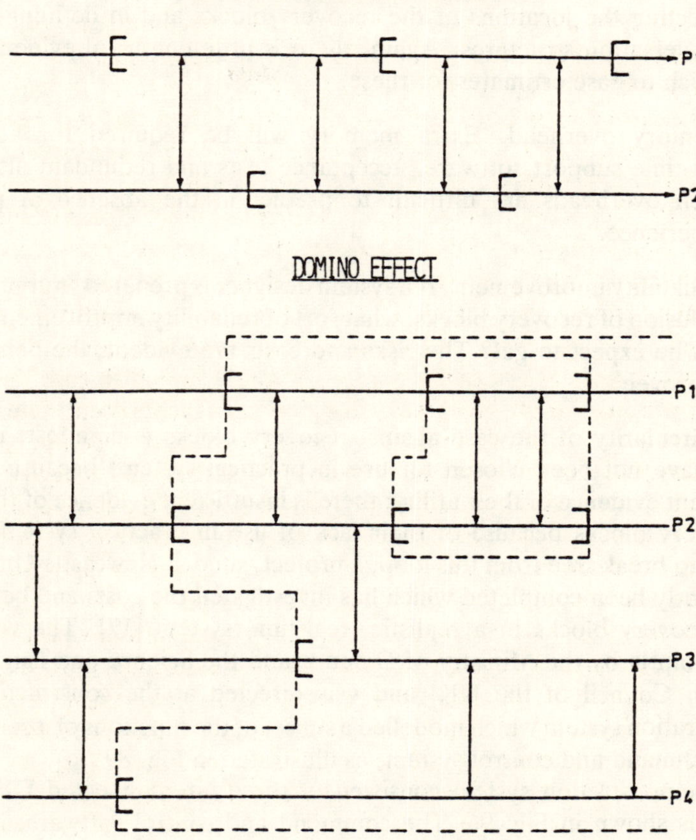

Fig. 95

(iv) Location of recovery blocks. For the effective utilisation of redundancy, recovery blocks should be used in those sections of the software most likely to contain faults which would cause system failure. The unpredictable nature of software faults makes this task extremely difficult.

(v) Development cost overhead. Software development overheads resulting from the use of recovery blocks can be divided into a fixed part and a proportional part. The fixed part will arise from the need to provide additional run-time environment software to support the operation of recovery blocks and conversations. The absence of a standard environment to provide this facility adds significant cost risk for any project contemplating the use of recovery blocks. The proportional part of the cost will be derived from the design and implementation of acceptance tests and redundant alternates, but will also include effort associated with selecting the locations of the recovery blocks and in defining suitable conversation structures. Again, there is little empirical evidence upon which to base estimates for these.

(vi) Memory overhead. Extra memory will be required for additional run-time support software, acceptance tests and redundant alternates. Such overheads are difficult to predict in the absence of practical experience.

(vii) Reliability improvement. If a system designer is prepared to argue for the inclusion of recovery blocks, what sort of reliability improvement, if any, can he expect to get? The risks and costs are evident; the benefits are unproven.

The circularity of the case against recovery blocks is manifest: recovery blocks have not been chosen for use in practical systems because there is insufficient evidence of their utility; there is insufficient evidence of the utility of recovery blocks because of their lack of use in practical systems. In an attempt to break free from this loop, a project, sited at Newcastle University, has recently been completed which has investigated the costs and benefits of using recovery blocks in a realistic, real-time system [17]. The work was funded jointly by the Ministry of Defence and the Science and Engineering Research Council of the UK, and was directed at the construction of a demonstration system which modelled a subset of the functions of a centralised naval command and control system, as illustrated in Fig. 89.

The demonstration system consisted of three interconnected DEC computers, as shown in Fig. 96. The command and control software, in which recovery blocks were included, was written in CORAL and based on MASCOT [11]. This ran on a DEC PDP 11/45, to which was connected the recovery cache described above, and a command console via which an operator could invoke command and control functions. The command and control machine was connected, by a parallel link, to a Unix-based PDP 11/45. This acted as a file-server on which monitoring output from the command and control machine was logged. The actions of own ship's sensors and weapons were simulated by MASCOT/CORAL software running on an LSI/11. Simulation scenarios were stored on the file server and read via a serial link. A graphics console was provided to allow an operator to control the operation of the simulator and to display the current state of the simulation. Communication between the command and control software and the simulated weapons and sensors was achieved via messages passed across a serial link.

The functionality of the command and control software was based upon anti-submarine warfare scenarios in which an operator would guide a torpedo-carrying helicopter to engage a hostile submarine. The command and control software was constructed in such a way that the software fault tolerance embedded within it could be either enabled or disabled. By running the command and control software in these two modes for various scenarios, comparative overall MTBFs could be obtained. Moreover, by using the monitor output from the command and control software, the fault coverage provided by the software fault tolerance could be estimated by determining the number of potential failures which were averted.

An important aspect of the work was the development of a scheme to apply the conversation principle to MASCOT software: a set of concurrent processes, termed activities, which interact through Inter-activity communication Data Areas (IDAs). The approach adopted was to define, at system construction time, static conversation structures called 'dialogues' [18]. Each dialogue was created with a unique name, nest level (since dialogues, like conversations, may be nested), activity list (to define those activities which are permitted to use the dialogue) and IDA list (those IDAs via which dialogue activity members are allowed to communicate). Each activity is created with a set of dialogues which it may use; dialogues may be entered or exited and this is essentially the way an activity establishes and discards a recovery point explicitly. A recovery block called by an activity will be passed the dialogue name to be used when establishing the recovery point of the block.

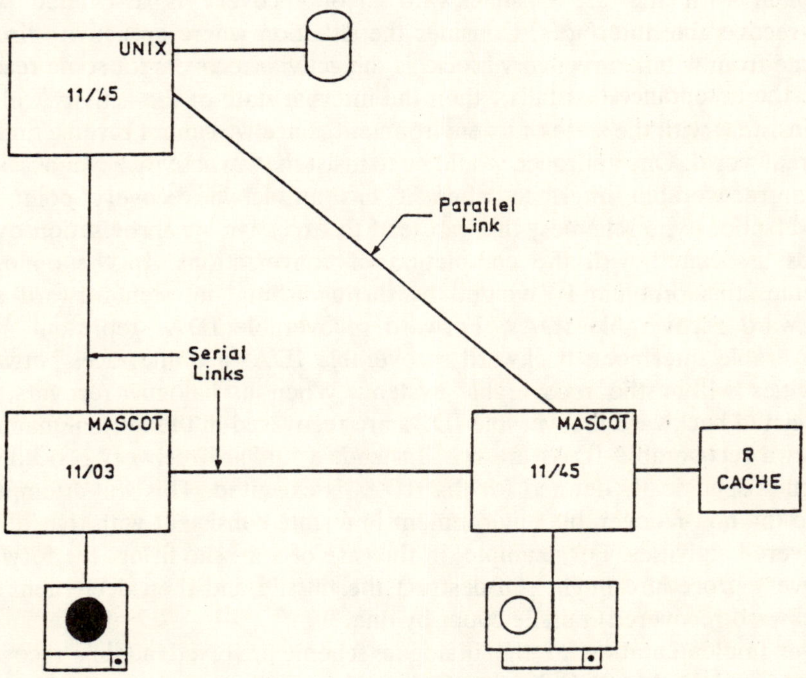

Fig. 96

EXPERIMENTAL OBSERVATIONS AND RESULTS

1 STATIC FORM OF CONVERSATION, TERMED A *DIALOGUE*, DEVISED FOR MASCOT

2 UTILITY OF DIALOGUES AND RECOVERY BLOCKS DEMONSTRATED

3 RELIABILITY IMPROVEMENTS
70% OF POTENTIAL FAILURES AVERTED
135% INCREASE IN MTBF

4 COSTS AND OVERHEADS
60% ADDITIONAL DESIGN & CODING EFFORT
33% EXTRA CODE
35% EXTRA DATA MEMORY
40% ADDITIONAL RUN-TIME
(30% SYNCHRONISATION OVERHEAD)

Fig. 97

The principle of static, named, dialogue structures is important since it provides good design visibility of the intended recovery structure. One major problem with the use of backward error recovery is associated with non-recoverable interfaces. Consider the situation where an activity fires a missile from within a recovery block. If the activity recovers for some reason (e.g. the acceptance test fails), then the internal state of the activity will be inconsistent with the state of its environment, since we cannot reverse time in the real world. One approach would be to insist that an activity never accesses a non-recoverable interface when it has an active recovery point. In conversation-type schemes, this can lead to excessive synchronisation overheads associated with the completion of conversations. In the dialogue scheme, this problem is avoided by distinguishing between forward and backward recoverable IDAs. Forward recoverable IDAs represent non-recoverable interfaces; backward recoverable IDAs are interfaces between activities within the recoverable system. When a dialogue recovers, all associated backward recoverable IDAs are recovered in the normal manner; forward recoverable IDAs are not. Instead, a forward recovery procedure, which is specifically defined for that IDA, is executed. This will attempt to place the non-recoverable environment in a state consistent with that of the recovered activities. For example, in the case of a missile firing, the forward recovery procedure might self-destruct the missile and then decrement the (backward recovered) missile count by one.

The implementation of the dialogue scheme involved adding recovery software to the MASCOT run-time kernel to support recovery blocks and dialogues, and enhancing the recovery cache software to accommodate concurrent MASCOT activities. Some 3000 man-hours of effort was expended in this work and the MASCOT run-time kernel size was increased by approximately 25%.

The results of reliability measurements on the demonstrator system [17] indicated that approximately 70% of software failures were averted by the use of software fault tolerance and the MTBF increased by about 135%. In fact, around 90% of all command and control software faults were successfully detected but hardware faults in the prototype recovery cache, and residual bugs in the recovery software of the MASCOT kernel, prevented successful recovery. In the absence of such deficiencies (which one would expect for standard, re-usable hardware and software), an increase in MTBF of 900% was predicted.

The price paid for this increase in reliability was as follows [17].

(i) 60% increase in the cost of developing the command and control applications software;

(ii) 33% extra applications code was produced;

(iii) 35% extra applications data memory was required;

(iv) 40% additional run-time was required (30% dialogue synchronisation, 8% cache bus monitoring).

6. Conclusions

Since the 1950s, fault tolerance has been used to improve the reliability of hardware systems. The reducing cost of hardware and the increasing functionality of integrated circuit devices has led to the development of fault tolerant multi-processor and local area network systems where protective redundancy is applied at the processor and computer level, respectively. The inclusion of redundant computers in a local area network is particularly attractive in military applications since the geographical separation of the redundancy can lead to a system which is tolerant to both operational faults and action damage. This paper has described the essential features of a dynamically reconfigurable, local area network, called ADNET, which has been specifically designed to exploit these potential benefits for a distributed naval command and control application.

Traditionally, fault tolerance schemes have only considered the physical failure of hardware components, although it is often the case that computer system failures are the result of residual software design faults. Various software fault tolerance techniques have been proposed during the last decade but there has been little evidence of their widespread use in practical systems. However, an experimental system, recently constructed at Newcastle University, has demonstrated that software fault tolerance can significantly increase the reliability of real-time software, and an account of this work has been included in this paper.

The increasing complexity of hardware systems, and in particular the advent of VLSI devices of customised design, suggests that conventional assumptions regarding the absence of hardware design faults in systems can no longer be considered as generally valid. Consequently, it is likely that fault tolerant systems of the future will require redundant components of independent design to be added to hardware systems, in a similar manner to that currently proposed for software systems. Inevitably, the increasing ease with which we can implement computer systems exposes our inability to specify and design them correctly and, in the presence of such imperfection, we must become more tolerant!

References

1. ANDERSON, T. and LEE, P. A.: 'Fault tolerance: Principles and Practice', Prentice Hall, 1981.
2. BOURICIUS W. G. *et al.*: 'Reliability Modelling Techniques for Fault Tolerant Computers', IEEE Transactions on Computers, C-20(11), pp. 1306–1311, 1971.
3. KEILLER, P. A., LITTLEWOOD, B., MILLER, D. R. and SOFER, A.: 'On the Quality of Software Reliability prediction', proc. NATO Advanced Study Institute on Electronic Systems Effectiveness and Life-Cycle Costing, Norwich, UK, 1982.
4. SHORT, R. A.: 'The Attainment of Reliable Digital Systems Through the Use of Redundancy – A Survey', IEEE Computer Group News 2(2), pp. 2–17, 1968.
5. WENSLEY, J. H. et al.: 'SIFT: Design and Analysis of a Fault-Tolerant Computer for Aircraft Control', Proc. IEEE 66(10), pp. 1240–1255, 1978.
6. HOPKINS, A. L., SMITH, T. B. and LALA, J. H.: 'FTMP – A Highly Reliable Fault-Tolerant Multiprocessor for Aircraft', Proc. IEEE 66(10), pp. 1221–1240, 1978.
7. REPTON, C. S.: 'Reliability Assurance for System 250, A Reliable, Real-Time Control System', First International Conference on Computer Communications, Washington (DC), pp. 297–305, 1972.
8. KATSUKI, D. *et al.*: 'Pluribus – An Operational Fault-Tolerant Multiprocessor', Proc. IEEE 66(10), p. 1146–1159, 1978.
9. GASDEN, J. A.: 'ADNET: An Experiment in Computer Networks for the Royal Navy', Proc. 3rd. Intrnational Conference on Distributed Computing Systems, 1982.
10. HILL, J. S. and STAINSBY, M. G.: 'A Highway for Intercomputer Communication', Journal of Naval Science, 6, 216, 1980.
11. MASCOT Suppliers Association: 'The Official Handbook of MASCOT', RSRE, Malvern, UK, 1980.
12. LAKIN, W. L. and MOULDING, M. R.: 'The ADNET Communications System: Inter-Process Communication in a Fault Tolerant Local Network', Proc. Third IFAC/IFIP Workshop on Achieving Safe Real-Time Computer Systems, pp. 233–238, Cambridge, UK, 1983.
13. TILLMAN, P. R.: 'ADDAM: ASWE Distributed Database Management System', Proc. 2nd. International Symposium on Distributed Database Management Systems, North Holland Publishing Company, 1982.
14. LEE, P. A., GHANI, N. and HERON, K.: 'A Recovery Cache for the PDP-11', IEEE Transactions on Computers, C-29(6), p. 546–549, 1980.

15. RANDELL, B.: 'System Structuring for Software Fault Tolerance', IEEE Transactions on Software Engineering, SE-1(2), pp. 220–232, 1975.
16. CHEN, L. and AVIZIENIS, A.: 'N-Version Programming: A Fault-Tolerance Approach to Reliability of Software Operation', Digest of FTCS-8, Toulouse, pp. 3–9, 1978.
17. ANDERSON, T., BARRETT, P. A., HALLIWELL, D. N. and MOULDING, M. R.: 'An Evaluation of Software Fault Tolerance in a Practical System', to appear in Digest of FTCS-15, Ann Arbor, 1985.
18. ANDERSON, T. and MOULDING, M. R.: 'Dialogues for Recovery Coordination in Concurrent Systems', In Preparation.

Index